Anonymous

Matter, Force and Spirit

Scientific Evidence of a Supreme Intelligence

Anonymous

Matter, Force and Spirit
Scientific Evidence of a Supreme Intelligence

ISBN/EAN: 9783337417420

Printed in Europe, USA, Canada, Australia, Japan

Cover: Foto ©berggeist007 / pixelio.de

More available books at **www.hansebooks.com**

MATTER, FORCE, AND SPIRIT

OR

SCIENTIFIC EVIDENCE OF A SUPREME INTELLIGENCE

" Faith were science, now,
  Would she but lay her bow and arrows by,
  And arm Her with the weapons of the time."

G. P. PUTNAM'S SONS

NEW YORK   LONDON
27 WEST TWENTY-THIRD STREET   24 BEDFORD STREET, STRAND

The Knickerbocker Press

1895

Electrotyped, Printed and Bound by

The Knickerbocker Press, New York

G. P. Putnam's Sons

PREFACE.

IT is believed that the advance of science has developed fundamental truths which strengthen the hands of faith. To briefly point out some of these is the purpose of this little volume.

The well-informed thinker knows that so far as determined by pure reason, the question of the existence of God is one of probability only. It is only by the accumulation of truth that learning comes to the aid of faith and hope.

In the contemplation of Nature, "Such is the order, fitness, and beauty, whether in the infinity of space, or in its unlimited division, that even with our little knowledge of things, all language loses its vigor, and all numbers their power of reasoning. Our judgment is lost in speechless eloquent astonishment."

By unavoidable convictions of cause, we complete a regression from effect to cause to a Being unconditionally necessary to such a world of life. Yet in this mental process, as in every kind of so-called proof of the existence of God, derived from speculative reason alone, we ascend from definite experience to a highest Cause existing outside of experience; to a conclusion of necessary existence, from given existence; and of course all such reasoning is fallacious and indeterminate. God is thereby disassociated from matter and force, and made a prior existence and a Creator; whereas matter, force, and God are co-existent and co-eternal. The very idea of God as spirit is inseparable from the idea of matter; and without such association He is no more conceivable than is a God without time, or space.

This concept of co-existence excludes the idea of the creation of matter, or of force, and presents the universe as an eternal reflection of spirit; matter and force being merely external expressions of Divine Self.

Conceding such Divine existence, we concede His knowledge of all the possibilities of His own laws, and of their culmination in organic life and mental power. In short we concede His prevision of man's existence on earth. And unless we regard man as a Divine plaything, the disport of an hour, a creature given being and then expunged from being, we may reasonably conclude, that his earth-life is a prelude to another life upon which he enters with the product of his earth knowledge and of his virtue.

CONTENTS.

CHAPTER I.

CHAPTER II.

CHAPTER III.

CHAPTER IV.

CHAPTER V.

CHAPTER VI.

CHAPTER VII.

CHAPTER VIII.

MATTER, FORCE, AND SPIRIT.

CHAPTER I.

INTRODUCTORY.

All Knowledge Relative—Ideas of Force and Matter Restricted—
Broadening Influence upon Religion of Science—They are in no
Sense Antagonistic—The too Aggressive Spirit of Science—The
Conservation of Energy—One Law for All Physical Phenomena
—Limit of the Inquiry.

BEFORE we concede to materialism that "the mind of man is potential in the sun," let us be sure that we understand matter and force,—so far, at least, as to exhaust what experience and the analysis of these subjects teach.

Time, space, matter, motion, and force, in their largest sense, are eternal. The contrary is inconceivable. Neither their creation nor destruction are thinkable. If they exist of *necessity*, they have always existed. And whoever asserts *the creation* of matter by a Supreme pre-existent Intelligence, asserts that during an eternity of time matter was not ; and that at some moment Supreme Wisdom changed,—which is inconsistent.

Matter is the matrix of man's education. It is the basis of his physical action and conceptions, and his whole mentality is founded therein. From it he derives his ideas of spirit, and a part of his spiritual growth is a process through matter.

Knowledge begins in the unknown and ends in the unknowable. All that man has acquired relates to but a brief interval between limits, from a time past, to a not very distant time future. Before these limits, an eternity;—beyond them, an eternity, about which he knows nothing.

Though imagination and speculation are untrammelled by boundaries, all positive knowledge is confined within these adamantine walls;—a relative extent in time, and a relative extent in space. All knowledge is relative, and all thought relative. "The highest reach of human science is the scientific recognition of human ignorance." All that we know, are the forms and the laws of phenomena; and absolutely nothing as to the ultimate nature of things.

Experience, with collective points of resistance, called matter, invests us with the idea that they are possessed of inherent force. Grouped under certain conditions of environment, they move as if under a directing agency, which we call life, and there result animate or inanimate forms, and even intelligence.

But who can assert that, instead of a material universe composed of solid, ponderable, ultimate units of matter, indivisible and imperceptible—this is not a spiritual universe, with what we call matter only a means to an end? A material manifestation of spirit?—"spirit and matter holding each other in mystic unity and equilibrium." Of course this is not what we accept as most probable, but there is always possibility beyond actuality, or the seeming solid materiality of our surroundings. While true learning destroys the self-assurance of men, it is not credible that, with its extension, the world is more sceptical; that as the supreme limits of knowledge are approached, there is an increase of atheism. Investigation has dispelled illusion. Experiment has established law. Discovery has destroyed superstition. Science has made the path to God clearer, plainer, and more attractively beautiful, instead of more obscure. The dense undergrowth of former ignorance with its forbidding gloom and mystery has been swept aside, and the lovely sunlight of truth illumines the way everywhere. Instead of many gods, mankind has been led to one God; instead of spirits of all degrees, they have been shown the one all-pervading Spirit, the Inspirer and Controller of Nature. The advance of science has not favored the material side only, for while it has given more direct and more simple views of natural phenomena, it has as steadily pointed to a universal Spirit.

Religion in its broad sense, has nothing to fear from science; on the contrary, it should hail with delight every scientific discovery, every truth established, not only as an approach to the unity and simplicity of fundamental law, and to emancipation from the blind assertions of the sceptical as to mysterious possibilities of Nature's forces during æons of time, but as conducting us to loftier views of the material universe. There are those ready to assert that matter has a manifold variety of physical, inherent powers, quite sufficient to account for the life of plants, of animals, and of men; that life is potential in the sun; and that it may be manufactured in the laboratory of the chemist. Because equivalents of energy are traced in the thought action of the brain, and analogies to polarities of force are seen in nerve life, they are quite ready to deliver themselves and their beliefs over to the possibilities of the mysterious forces of nature, which are declared to be all-sufficient though too elusive to comprehend. It was long doctrinally held that each of the physical affections known as light, heat, electricity, magnetism, and chemical affinity, was due to the presence of a peculiar force inherent in the final units of matter, whose characteristic was correspondent with the particular manifestation or mode of motion set up in each form of phenomena. So long as it was maintained that matter was thus endowed with such a multiplicity of properties, so numerous, co-resident, yet distinct, there was tacitly conceded to it a capacity almost unlimited. There is no need of surprise, in the presence of such concessions to the mysterious, that imagination should dominate, or that wonder should replace reason. In these assumptions, there was a surrender of the key position either to unlimited credulity, or to arrogant assertions that matter alone is all-sufficient to form life, and that man need not look beyond it for a God.

The sooner such assertions are stripped of this mask of the mysterious, the more advantageous to the scientific and religious spirit of the age. The truth then will be the more clearly seen, that though the natural energy of matter suffices for its aggregation into earth forms, and for the preparation of conditions favorable to life, it is utterly impotent to confer life itself; and that matter is only a domino veiling universal life. The true spirit of patient, reverent, and truth-seeking science, is not in sympathy with confident assumption. Discovery has followed discovery, until the very confines of inquiry in some directions have been reached. Light and heat have been rescued from the domain of entities, and are, like sound, regarded as modes of motion. Chemical action is wholly separated from the mystic term "chemical affinity"; its polar energies have been shown, and that its interchanges are between electro-positive and electro-negative elements. Magnetic and electric forces are no longer thought to be fluids—mysterious forms of matter, eccentric accumulations in space of wondrous powers and capacity; but merely vibrant conditions of substance, or modes of its motion, belonging to all matter, but more manifest in some than in other elements; in some, affecting more the surfaces, in others, the entire mass of bodies. Its interchanges are waves of motion, impelled by higher intensities, or by greater volume of the released energies of one portion of the body, or of the surface, over those of the other portion. From time to time, glimpses from scientific standpoints have suggested the existence

of a single primary energy underlying all processes
of nature; but the direct view was not attainable
until the truth of the conservation of energy, or the
persistence of force had been established; then it
followed, that the different phenomena of light,
heat, electricity, magnetism, sound, chemical power,
etc., are rendered possible by the transference of
equivalent force from one of these forms to another.
Of course, under this view, all phenomena become
merely relations of succession.

There is no truth in nature more fundamental
than the principle of the persistence of force; where
force seems to have vanished, it has merely taken
some other compensating form.   Formerly, the annihilation, or the sudden creation of force, without
apparent cause, was considered in thought perfectly
natural, and no more singular, than that heavy
bodies should fall; though why they were heavy, it
seems never to have occurred to anyone before
Newton, to inquire.   It is impossible to over-esti-
mate the importance of *a law referring to a single
origin, all physical phenomena.*   Beyond the approximate scientific unity, there follow homogeneousness
of doctrine, and unity of philosophical method.   In
unity of credence is the beginning of harmony;
therefrom flows the useful, the perfection of equity,
and there begins universal benevolence.

Beyond the impetus and accord thus given to
truth, are increased facilities for the study of abstract
knowledge in its broadest conception.   Its pursuit
is man's most exalted function.   It alone enlightens
him in his relations to his race, and to nature, and

best fits him for his position at her head. The inquiry can never be wholly removed from the metaphysical aspect, since positive demonstration is impracticable. Matter is known only through motion, and force only through matter; and there will always remain a possibility, beyond the real and useful probability. But can Faith, or hope of a future life, claim anything more? If spirit be not the source and energy of all things, the evolving principle of all life in that which we call matter, then matter, dense, helpless, and inert, and having an equally helpless force,—is the soul and source of all. Life is then a hideous, grinning satire, a burlesque upon effort and knowledge,—objectless and purposeless; and Nature nothing more than a vast, blind, grinding engine, unbidden and unknowing, working in an aimless and spiritless universe.

It is proposed to examine briefly that knowledge, which comes of experiment and experience, called science; to see how far, if at all, it antagonizes spirit; to view there the horizon of the sceptic, and, perchance to get a better spiritual outlook. We may then determine whether such knowledge conducts to a lower or a loftier ideal; nearer to, or farther from a universal and supreme spirit.

CHAPTER II.

MATTER.

I T is our purpose to treat matter and force from a purely physical or material standpoint, viewing force according to the mechanical understanding of force, as a quantitative energy; and stripping matter of its mysterious asserted entities, and so far as may be, laying it bare by analysis, that its possibilities, and intrenched life-creative powers may be estimated. We therefore assume matter to be a reality, and not collections of immaterial points of resistance, and if its masses be real, that their combining units are also real. An old and still contested question is of their unlimited divisibility, and it has grown to factitious proportions by the extension to abstract possibility, of the likeness of actual process. It is of no importance that we may abstractedly extend the division of what are the real final units of matter; there is a vast difference between the conception of successive division, indefinitely prolonged, and the actual limit of divisibility.

The real question is, do the facts of nature indicate a limit to divisibility? Though this may not be the limit of conception, yet it is the limit of our concern with the inquiry, for it is the limit of the operation of law through any period of duration. This, presumptively, has been eternal, and consequently, so far as we are concerned, any ultimate consequence has already been reached. To say that matter is every thing visible and tangible, is incompletely defining it. To say that it is substance, of which all things are composed, is to say that matter is matter. The only possible standard of thought is the ultimate conception. The instant we admit matter to be real, and distinguished from force, its ultimate measure, we admit its final units to occupy in space the three dimensions, of length, breadth, and thickness, and that they are incompressible; otherwise they are not final, for whatever is compressible is made up of parts, and compressibility is the capacity of diminishing interstitial spaces. Hence the reality of matter, to which we arrive in conception, as our ultimate thought, is that of final units of substance, of three dimensions without constituent parts, and wholly incompressible and impenetrable, opaque, colorless, and ponderable and having absolute hardness. A homogeneous unity, neither disruptured by shock of impact, nor divided by friction, or other influences. The idea of the composition of matter from indivisible ultimates, is a growth from scientific thought. At first it was a vague metaphysical notion, originating with Leucippus, but under the bold hand of Dalton, it became a brilliant and fertile reality.

The deductions of chemical analysis and synthesis evince that the practical divisions of matter are soon reached. The balance of the chemist has also overturned the belief so long entertained, of the destructibility of matter. Now the conception of its diminution or expulsion from existence, is as impossible as that of its increase, or appearance from nothing. And as each of its final units is the matrix of an inherent energy, and as each also represents by its never ending motion, a mechanical force, the idea of the creation, or annihilation of matter, obliterates at once all idea of laws of force. It will therefore be accepted, that the quantity of matter, and therefore of inherent energy in the universe, is always the same.

The figure of the ultimate units of matter, though unchangeable, is not practically determinate, the inquiry, for obvious reasons, being removed from demonstration. The inherent force of each unit tending to uniformity of distribution, and concentrating on itself would—granting mobility—produce a sphere. But absolutely fixed and incompressible as are these units, any resultant shape consequent upon any conceivable force, is unthinkable. This is true of all conjectures whatever upon their shape, as proceeding from a constraining force. And yet the subject is not wholly removed from the domain of reason. The course of nature in all of her manifestations is a procession of uniform simplicity, and all experiences irresistibly shape ideas to the belief in one comprehensive method.

In minute, component forces, before polar inter-

changes, or other causes, have forced a secondary
arrangement of units, we arrive at a sphere as the
representative figure. In semi-fluidity, where vibratory action predominates over attraction, there is in
small aggregations of matter a spherical resultant.
And cosmical masses, under the free action of central forces, are rolled into spheres throughout the
heavens. As, therefore, all discernible forms tend
to the spherical, in this uniformity there is involun-
tarily suggested the idea of an ultimate spherical
archetype of form, justified further by the very
hypothesis of the investiture of matter with inherent
force. For as the final units of matter are homogeneous solids, their motions would be prejudiced in
particular directions, unless the inherent force of
each was so distributed, as to be capable of acting
with equal energy from any point of the surface
upon environing matter ; and no figure other than the
sphere admits of these conditions. It is a figure of
equilibrium, in being a figure of equal distribution
of its own inherent energy. And since in nature,
there is no warrant of predisposed direction of energy, the burden of evidence renders it probable
that the figure of the final units of all matter is
spherical ; and if this be the case, it necessarily
follows, that the natural figure of all multiple
atoms, or molecules, whose units are of the same
magnitude, are also spherical. Another question is,
whether or not there is uniformity of magnitude in
the final units of matter. If the ultimate nature of
all matter, in its final units, be the same, this question
must be answered in the negative ; for then there is

no other way of accounting for the apparent differences of so-called elements, than that there is a difference of magnitude between the final units of different elements. As to whether there can be radical differences of elemental matter, depends upon the nature of the inherent force of the final units of each element. If these forces be, or can be, different, then, of course, each element in its mass presents corresponding properties to these inherent forces. This inquiry will be fully discussed in the chapter on force, but for the present, it will be assumed that matter is identical in its final divisions; and of course, that each final unit of all forms of matter is invested with precisely the same kind of force. Under this assumption, we shall examine the evidence as to the difference of magnitude of the various forms of elemental matter. All composition and resolution of matter come from the action of forces,—those of mechanical motion, and those inherent in the final units; the last, the assembling, and the first, the disintegrating; the two together sift matter. Under the dominion of these forces, matter is in unceasing motion, not only the consentaneous motion of the units of the mass, but the vibratory motion of those units; for none of the physical affections of matter, known as light, heat, electricity, magnetism, chemical action, etc., are admissible, without a conceded motion of the ultimate units of matter, with interspaces for that motion. Equal diffusion of motion throughout any mass is consistent only with uniformity of resistance, and a uniformity of the forces acting in the mass.

It is upon the equal diffusion of motion, through the composing units, and the uniformity of their resistance to motion, that depends the stability of *elemental* divisions of matter.

If the units are dissimilar, their degrees of displacement are unequal, their motions neither consonant nor reciprocal, and they will consequently be segregated, under the action of a continuous incident force, and reunited into groups of like units, the arrangement being the more stable, as the units are more alike. Further, if the units of all elements be of equal magnitude, the degree of their juxtaposition, or the density of all elements under the same environment should be the same, which is not the case. Proofs and arguments could be readily multiplied to the same end, drawn from the laws of chemistry,—from considerations of heat and other sources, but they would be out of place in a work of this nature. Suffice it to say, that we must either admit differences of magnitude of the final units of the different elemental forms of matter; or that the final units of these elements are invested with inherent forces in each element, differing from those of any other element. Now as there are in nature about seventy known elements, metallic and non-metallic, it follows that there must be a corresponding number of different inherent forces, a particular kind for the final units of each element. This is not only absurd, but impossible, as will be shown in the discussion of the subject of force. But supposing all these points settled, are they decided with reference to our earth alone? If so they are special.

Or are questions of divisibility, indestructibility, form, magnitude, etc., of final units, to be confined to our solar system? If so they are still special.

Astronomers no longer grope in the heavens to determine problems, or to verify evidence limited in its scope, or peculiar in its application; they seek generalities and the broadest types. From general laws inquiry extends to more fundamental laws in all channels of knowledge, so impressed is the scientific spirit, that the matter and energy of our system are connate with those of all others, a unity, a oneness, a likeness of universality. This is an increase of conviction through the multiplication of experiences,—experiences, the researches of all men, adopted as individual experiences, and again extended, impossible either to controvert or to avoid. To declare that matter is of a definite limit, though contained in an indefinite extension; and that particular elemental forms of it are limited to particular parts of space;—or that matter conformable to one part of space is unconformable to other portions;— is to oppose a development of thought now become universal. It is to retrograde amidst general progression. It is to turn toward the crude opinions of man's primitive state, when, in his consummate vanity, he regarded himself as the centre of the universe and its final purpose ; the pride of the earth and the admired of heaven ; when he believed that the sun, moon, and all the stars of the celestial sphere made daily obeisance around his little planet. And when in his conceit he complacently believed that angels ministered to him, and

in his supreme arrogance scorned to think himself
subject alike with all things to nature's laws ; when in
fine, he surveyed himself as the exclusive and unre-
mitting object of Divine attention. But knowledge
has conducted us to wider, nobler views, not less
reverent, but more humble.

Space is infinite. The contrary is not conceivable. If then but a part of space contains matter,
the infinite part beyond is a useless void, and any
definite part, however vast, is simply a finger's span
to the infinite part. This idea of space, and the
occupancy of a mere crib-work of the whole, is
utterly inconsistent with any natural conception of
an Omnipotent and Omnipresent Being. For He is
all-comprehensive, all-embracing, and His correspondent and parallel can be found only in an infinite universality. Wisdom, space, force, matter, no
one of these could be co-equal in completeness if
any were limited ; for then since the Infinite is omnipresent, absolute wisdom and power would to no
purpose extend to an infinite empty space.

If that infinite space be abandoned of God, of
matter, and of force, He is not omnipresent, and
there remains an infinite uncontrolled region where
He is not. Nor is He infinite, if not omnipresent.

Now let it be supposed for a moment, that there
is no infinite co-existent intelligence, whose supreme
wisdom directs all things ; that there is merely a
self-existence from necessity, which always was, and
always will be. There is then *of necessity*, space ;
and *of necessity*, matter. If then matter fill only
a mere patch of space, necessity has only a partial

action ; it is neither uniform, nor universal, and this is not necessity.

What *necessarily* exists, universally exists. Whatever is *a necessity* to a part of space, is a necessity to all space. Whether then, from intelligent omnipotence, or from necessity, the universe exists, as it exists, matter must be regarded as the universal feature of infinite space.

But to return to actual discovery. Telescopes of increasing range are filled with increasing fields of matter, as dense, as vast, in consolidated or diffused magnitude, as that nearer and better known, and at distances so great, that the mind wanders in their contemplation. And the assumption of unlimited worlds, and world forms in the infinite beyond, is defensible by analogy. But is the matter of this immeasurable space the same as that surrounding us? In the subject of force it will be shown, that as far as is possible to determine, the laws of gravitation govern all matter and stellar systems,—and that there is a consonant parallelism in that portion of the universe within our view.

Physical science persuades us that our sidereal system is but an atomic agglomeration from many others. The central forces of attraction have forced into collision vast worlds of consolidated matter;— the shock of impact has transformed all their progressive motion into the vibratory motion of heat, dispersing and expending this matter to amazing limits ; diffusing it into its final units. Under the influence of this expansive power the matter of our system has passed into other spheres,—and that of other spheres undergoing similar experiences, has commingled with our own.

Aërolites of solid matter, identical with terrestrial matter, are shot from stellar regions to our earth; and wandering comets are driven into our sun, or pass in eccentric orbits forever beyond its influence, to other solar systems. This perpetual intermixture, without beginning or ending, must have produced correspondence, in fact, undistinguishable sameness everywhere. The spectroscope decides the presence in remote worlds, of elements, metallic and non-metallic, identical with those of ours; and a fair analogy warrants us in extending the congruence indefinitely; and in the conclusion, that there is an infinite extent of matter in no respect different from that occupying our solar system.

Through endless cycles it has pursued a common course, and through unceasing duration will continue to repeat its changes of aggregation and segregation. The telescope in its scrutiny of the universe will find no void short of infinity, and the microscope, no element different from ours, or younger than eternity. Another question is the identity of matter.

Do its diverse elemental properties arise from inherent differences of the force of the final units of matter of the different elements, or do they arise from differences in the action of the same inherent energy, by reason of the differences of magnitude of the final units?

In other words, are ultimate units possessed of primitive forces, intrinsically different, or are the forces of all final units precisely the same? The

complex matter of organic compounds, distributed
in pleasing pictures, and in lavish profusion around
us, are appearances of variety and diversity; yet in
a few chief, and some subordinate elements, by the
varying of the constituent proportions, are com-
prised not only all animal and vegetable substance,
but the great bulk of terrestrial matter. It is commonplace knowledge, that these beautiful forms
furnish to the senses no evidence whatever of their
ingredients, and that the true standard of their
determination is force. Followed back to their constituent elements, in their decomposition, the decep-
tion of appearances, and not unfrequently to a
startling extent, of properties, continues. And
rigorous analysis is often necessary to distinguish
matter, in many states identical in appearance, yet
physically different.

On the other hand, so utterly unlike are the forms
assumed by the same element, that its equivalent of
force established by repeated experiment is the only
final umpire. Let us illustrate by a single example,
how color, odor, form, density, and other properties,
may arise from mere differences of arrangement of
constituent parts of matter. Under three states
wholly unlike, is the simple element, carbon. In the
diamond, it is transparent, and the hardest of all
substances, refracting light in a high degree; in
graphite, or black lead, opaque, black, quasi-metallic;
in charcoal, velvety, soft and porous; yet isolate the
ultimate units of each of these forms,—dissever
them from collective motion, and they are the same.

Again, dissimilar elements, by the arrangement of

their parts, may closely assimilate. Crystals of pure boron are brilliant, transparent, and octahedral; like the diamond of carbon, nearly colorless, and differing but little from it, in hardness and refractive power.

The reference of last resort, then, in questions of the identity of matter, is not to be found in mere mechanical properties, since it is through the elimination of these properties, that we approach the individuality of matter; ultimate units can have no essential difference not founded upon inbeing forces. But habits of thought and association of ideas have fastened upon us beyond eradication, ideas of matter, so that we unconsciously receive, as true expressions of it, mere symbols, or reflections to us of its outward form. Nothing can dispossess us of this delusion of thought, but an intimate knowledge of nature.

These perpetual repetitions are prototypes from which we fashion our notions of the actual, and we fancy that the differences between ultimate units, are as radical and intrinsic as are represented in the bodies themselves.

These illusions have even seized upon and infected the spirit of science. It has in this manner come to be supposed in the general mind—and science is not free from the error,—that in the disintegration of matter, each minute component carries along with it the continuation of all the properties that characterized the mass. But this fallacy is seriously shaken by the single reflection, that all the physical properties of matter with which we are acquainted, arising from forces, with the single exception of atomic energy in chemical action, are due to combinations of units into masses; and that they have no existence without such concretion, since they are modes of motion of many units, ceasing with the resolution of the mass into its constituent parts.

Therefore, in the dissection of matter to its final units, there are eliminated all those qualities, derived from their aggregated motion, and serving to distinguish the mass, as sound, light, heat, color, etc.

Stripped of physical differences thus arising, what are correspondences, and what is the diversity in matter remaining?

All final units being impenetrable, indestructible, incompressible solids, there is homogeneousness in all. They are equal in density, are ponderable, proportionate in gravity and inertia; all are non-elastic; there is no texture, color, or temperature; and there can be nothing left but *the asserted differences* in their inherent forces. These are the dual energies of each final unit, upon which existing doctrines base magnetic and electric phenomena; the attractive force upon which gravitation is explained; and the asserted distinct and separate, yet co-resident force of repulsion,—all united, it is alleged, in the same ultimate form of matter.

There is the highest probability that the interspaces of the final units of matter, in the densest mass, exceed, by many times, their diameters. If this were not the case, color and shade of surfaces would be impossible, as would also the temperature of bodies; for temperature is a measure of heat, and heat motion is vibratory motion of the mole-

cules, or multiple atoms of matter: therefore, temperature is the measure of vibratory motion of
matter, and as there are many hundreds of degrees
between the ordinary temperature of terrestrial
bodies, and the point which is called the absolute
zero of temperature,—so there must be a vast
amount of vibratory motion in all matter. Indeed
a wave of light forming a sunbeam, and passing
through glass, has been by Herschel, happily lik-
ened to a bird threading the mazes of a forest. One
must certainly emancipate himself from the delusive
notions of matter, as presented by the senses, and
from the chimera, that substance is intrinsically
what it seems to touch and vision.

Intervals of thousands, or even of millions of
times the diameters of the final units of matter may
not express their distances apart, yet through which
their individual power is so intense. There is really
nothing extraordinary in this, when the minuteness
of nature's processes are considered. Through the
microscope we are introduced to orders of life so
diminutive, that within the volume of the smallest
grain of sand, move millions of animated beings,
having phases of birth, growth, and reproduction;
and with organs of digestion, circulation, respiration,
and locomotion; and in every direction of inquiry,
appears a world of life, whose being begins where
the unaided perception of our senses ends.

It is important to our general plan, to outline very
briefly the nebular hypothesis of matter, which explains the formation under the laws of gravitation
of our solar system.

This hypothesis has become the common property

of science. And its probability has been fully established by mathematical and experimental demonstration. It plainly would be obtrusive to submit
these proofs here, as they are facts of physical science ; nor is it at all necessary, as we accept without
question this grand illustration of the law of gravitation, universal, so far as our visible universe per-
mits us to investigate.

The facts of our planetary system indicate that it
was once a connected mass with a uniform motion
of rotation. Otherwise it is impossible to explain
the common motion in the same direction of the
planets, both orbital and axial ; why the planes of
their orbits and those of their satellites and rings,
all nearly coincide ; why their orbits differ but little
from circles, and much besides. The theory supposes that loose masses of nebulous vapor, at first,
without specific form or movement, gradually assumed, by virtue of gravitation, a regular spheroidal
and rotating shape, lightest at the circumference,
and gradually increasing in density toward the centre, at which point the greatest density is reached.
It supposes that such was the original form of suns ;
that the substance of these, in this diffused state,
originally extended from their present condensed
solar spheres, beyond the outermost limits of the
orbits of the most remote of the planets which now
revolve about them ; and that by the combined processes of rotation and condensation, successive con-
centric rings were formed, beginning at the outermost
limits, which rings were finally broken and rolled
into spheres, becoming planets to the central body.

For instance, under the attractive force of all matter, the diffused vast nebulous mass would condense into a nebulous sphere, becoming constantly smaller, by which, according to mechanical laws, a motion of rotation, originally slow, would gradually become quicker and quicker. By the centrifugal force, which must act most energetically in the neighborhood of the equator of the nebulous sphere, masses would, from time to time, be torn away, and would continue their courses separate from the main mass, forming rings which subsequently became broken up, the matter composing them naturally agglomerating into spheres, single planets, similar to the great original sphere ; and by analogous processes of condensation and evolution of rings into planets, with satellites and rings, until finally the principal mass condensed itself into our sun. The following are, briefly stated, the principal physical facts from which the opinion is derived, that our own system has passed through the successive stages, from the nebulous to its present condition.

The earth is an oblate spheroid flattened at the poles. Proofs of its former semi-fluid state, and of its gradual contraction by cooling, are numerous and common. There is great reason for believing "that its shape is due, rather to a withdrawal through this process, of portions least subject to centrifugal force, or removed from the equatorial parts, than that the existing form is a modification of that, originally globular, by initial rotation. Since if the latter portions had, from a spherical shape, begun to be thrown outward, they could not have been restrained

within limits by any counter force, and the flattening
process must have continued indefinitely, had the
velocity been undiminished. The detachment of
annular rings is illustrated by those of Saturn. The
separation of parts, by inequality of motion, and
contraction, has its parallel in the recent rupture of
Biela's comet."

The planets exhibit regular graduations of densities, from those nearest, to those most remote from
the sun. Thus: "On the basis of mathematical
calculations, Mercury must be about the weight of
so much lead; Venus is nearly six times the weight
of so much water; the earth, as a whole, is four and
one half times the weight of water; Mars, a little
over three times the weight of water; Jupiter is a
small fraction over the weight of as much water;
Saturn is less than half that specific weight, or about
the weight of so much cork; and Herschel manifests
a corresponding decrease of density. This gradua-
tion is precisely what it should be, supposing that
they were all formed by the operation of a common
law, from an original sphere of fluid (gaseous) matter,
which must have been most dense near the centre,
and most rarefied at its outer surface." "There is a
similar relation between the distances of the different planets. Proceeding outwards and regarding
the asteroids as equivalent to one planet, each successive planet from Mercury is about double the
distance of the previous planet from the sun, thus
arguing their production from a common law."

"It might be supposed that after the evolution of
Mercury, the planet nearest the sun, there would still

be a residuum of nebulous matter surrounding the denser nucleus of the sun. Accordingly, we find an extensive mass of attenuated matter surrounding the sun, called the zodiacal light. And within this, yet more concentrated, and extending outward from the denser molten mass, is a fiery vapor, or incandescent atmosphere, enveloping the solar focus, and reaching from it to a vast distance." Now when it is remembered, that all nebulæ have one or more bright foci, upon whose movements the incandescent vapor is attendant, the analogy between them and our sun and system is strengthened.

Nor is this all. According to the principles by which periods of rotation maintain a relation to the mass of a given rotating body, it has been proved that the sidereal year of each planet actually corresponds to the period in which the sun must have rotated on his axis, supposing his mass to have extended to the orbits of such planets before they were thrown off. And the periods of rotation of the primary planets, with their mass in a state of vapor, extending to the orbits of their satellites, must in like manner have corresponded with the present orbital periods of those satellites.

Besides the foregoing, is Kirkwood's law; that " The square of the number of rotations of any given planet in its year, is to the square of the number of rotations of a second planet, as the cube of the diameter of the sphere of attraction of the first planet, is to the cube of the diameter of the sphere of attraction of the second planet. So definite is the relation between the forces and movements of the different planets, as to preclude all reasonable supposition that it came by chance." " The nebular theory answers to all the appearances of our system, and explains the motion of the planets, both primitive and present. It shows that the formation of the system has been successive, the most remote planets being the most ancient, and the satellites, the most modern. If from points of view like these, the stability of our system can scarcely be regarded as absolute, what it leads us to suspect is, that by the continuous resistance of the general interstellar medium, ether, our system must at length be reunited to the solar mass from which it came, till a new dilatation of this mass from the extreme heat shall occur in the immensity of time, and organize in the same way a new system to follow a similar career.

The process of re-uniting is, that under the resistance of the etherial medium, planetary rotations must become slower, the orbits smaller and less elliptical, and their periodic times shorter, until finally the planet plunges into the sun ; so that in a future too remote to be designated, all the bodies of our system will be re-joined to the solar mass from which they proceeded. All these prodigious alternations of destruction and renewal must take place without affecting yet more general phenomena : the mutual action of suns ; consequently these transformations of our system, too prodigious to be more than barely conceived of, can be only secondary, even local events, as compared to the far more extended, and even universal, transformations of consolidation and expansion of suns into suns, and those suns into other suns."

In culling promiscuously and from many sources, the abundant proof of this grand theory, our purpose has been to place it in the clearest manner before the reader, avoiding much that is complicated of proof and details: if the facts given be accepted, the probabilities are more than contestable, for the summary of evidence amounts to conviction. It fulfils all the conditions of a rational hypothesis, for it not only explains all phenomena involved, but it is the only hypothesis that will do so; besides it admits of unequivocal verification by experiment. It is the grandest summit from which we may view in its totality and endless perspective, the processes of Nature's fundamental law of gravitation. Alone, it is a wonderful vision; but it is more; for in its contemplation from the inchoate state and initial procedure of diffused matter to the composition of worlds, we see the simplicity and sublimity of Divine wisdom. In its light, our solar system vanishes to a mere point in space; its matter, to the consolidation of a vast annular ring cast off from some mighty sun, and whirled into globular form with possibly many others, a mingled multitude of primaries and satellites from the central orb, with whose volume they were once inter-diffused in a single nebulous mass. And within the depths of far more remote space, it may be, many billions of inter-solar spans, is still a vaster centre, from which this last mass with innumerable others was devolved, whose circumscription, though almost transcending conception, is yet

definite ; for in its turn it is an expansion and concentration of some greater approximation to the
indefinite beyond. Still there is measurement, circumference, cubicalness, and position of all these,
as positive and limited as for the smallest object
before us. We have the revolution of satellites
with planets around suns ; suns around greater suns ;
the greater suns around systems ; systems around
greater systems ; the greater systems around clusters ; and these around immeasurable zones, multiplied to incalculable immensity ; and yet beyond
their confines, there remains always an unattainable
extent, never to be compassed ; all agglomerations
rolled from nebulous diffusions,—each the production of an origin beyond. Yet in the infinity of matter and space, there is no absolute centre, for there
is no circumference ; hence no grand source can be
arrived at : nor are we in any sense concerned with
this question, only with that of interbalancing limits,
which probably space off here and there, at intervals,
the stellar universe. The cause of the nebulous expansion and diffusion of matter is, as has been stated,
heat, generated by the conversion of the motion of
masses through collision, into an equivalent motion
of vibration (or heat motion) of their units. In return, processes of condensation, the heat motion of
these units is parted with by cooling of the matter.
Helmholtz estimates that but one four hundred and
fifty-fourth part of the original mechanical force in
our planetary system remains ; but that the balance
converted into heat, would still be adequate to raise
a mass of water equal to the sun and planets taken

together,—not less than twenty-eight millions of degrees of the centigrade scale.

It is seen, therefore, that, by far the largest proportion of the original heat of our system, has been radiated into space in the cooling and contraction of the nebulous matter, to the present density of the planets and sun. The exterior boundary of the nebulous sphere was far beyond the outermost planet; and the curve of separation of the first nebular ring, where the centrifugal force, due to rotation, equalled the gravity, was greatly beyond the existing orbit of that planet. It is presumptive, then, that our solar system, in its successive processes of transformation of the aggregate motion of its planetary bodies into the molecular motion of heat, generated by plunging of the planets into the sun, and the loss of this heat by radiation into space, must in a remote future arrive at a state of entire consolidation.

Its orbital motion, which it now pursues around some distant centre, must in like manner cease by the ultimate integration and consolidation of its mass, and that of other systems, with that vast centre from which they, in common, originated. And further, if there be conceded an indefinite space, beyond stellar space, unoccupied except by an etherial medium, similar to that of the inter-planetary spaces, the progression of all stellar matter to final consolidation and rest, precisely analogous to that above traced for our own system, is inevitable, at some period, which, however distant, is not of infinite duration. For, however great the intervals of

time, and however great the masses moving to combination in one general aggregate, they must ultimately convert their motion of translation into
the molecular motion of heat, and by cooling, transfer this to the outer ether medium beyond them,
there to pursue its endless way.

It will be seen, that the foregoing conclusion is
reached, under the assumption that the space occupied by the stellar universe is definite, or circum-
scribed ; and that beyond it there is nothing to an
infinite distance, but the etherial medium. But we
see no good reason for departing from the conclusion
already reached in these pages, that space is indefinitely occupied by forms of matter beyond our vision, rigorously correspondent to those forms observed
by us ; that is that the stellar universe is of infinite
extent ; for the universe is not a dot, nor its amplitude
a disconformity. Under this aspect of matter, and
under the most extended application of the nebular
hypothesis, it may readily be shown, that all matter,
so far as we can comprehend the meaning of the
term " all "—will, in a similar manner, be transformed
into a mechanical state, where further change under
its own laws will be impossible. For let it be assumed (as will inevitably be the case) that suns with
their satellites, groups with their suns, clusters with
their groups, and zones with their clusters, should
concentrate upon a common centre of gravity.

The heat developed in this general integration of
masses so vast, moving to collision through intervals
almost immeasurable, under velocities constantly
accelerated, would rarefy their matter to an incalcu-

lable degree; and of necessity, the nebulous dispersion
following must overlap the orbits of stars immediately beyond those concerned in this aggregation.
" The resisting medium to the motion of such stars,
being thus augmented, greatly expedites their mutual
approach, under gravitation, to a centre ; and their
subsequent expansion, in turn disturbs, in like manner, the equilibrium, and hastens the concentration
of other systems. Therefore collisions throughout
space conspire to bring about more frequent collisions; and in continually extending the nebulous
matter, the inter-radiation must inevitably terminate
in the dispersion of all stellar matter whatever, into
the nebulous form, and the ultimate equilibration of
the heat of that matter." Here all further transformations must cease,—all operations of nature;
and the universe of matter would be doomed to
eternal changeless monotony, unless the Omnipotent
Controller of all should move within its depths the
resuscitation. Upon any assumption then, and
whichever way the subject be viewed, we arrive at
a period of the material equilibration, and eternal
changelessness of all matter of the universe. This
result is inevitable, under any of three possible suppositions: *i.e.*, whether we regard the number of
sidereal masses as limited, with an infinite ethereal
continuity beyond ; whether interstellar ether and
sidereal masses have limited boundaries, beyond
which, to infinity of extent, there is nothing; or
whether an infinite space be occupied, as is that
within the limits of our vision. There is a culmination, a decay, and a dissolution, inexorably following

upon the present view of the conservation of the force of the universe, and the operation of the laws of gravitation, which, so far as we are concerned, must be accepted as fully verified by the most extended observation and experiment; as much so, indeed, as any knowledge known to man. This unavoidable stagnation and practical death of the matter of the universe, in some distant future time, by no means infinite in duration, and brought about, as we have seen, by law, could in no other manner be reanimated to renewed form and to progressive life, than through an influence beyond law, and above it, and for which, law is insufficient. Here indeed in the material universal life, is a pleasing parallel to that terrestrial organic life, from which we are inseparable,—birth, growth, death. Now let it be remembered, that whatever has been possible under the operation of law, in the tendency of matter to any final state, has, during the eternal duration that has preceded us, been already consummated. And, as it has been demonstrated, that matter, under existing law, is steadily pursuing a course, toward either permanent and universal consolidation into one mass, or toward universal diffusion and equilibration, from either of which states there is no self-reaction; it necessarily follows, either that the fundamental laws of matter are not now, what they were at some anterior period of existence, or that matter has, at some former period, arrived at one of the states of stagnation and practical death already described; and from that state, has been restored, by Divine agency, to the condition of renewed life as we now see it in the starry heavens.

It has doubtless been observed that the foregoing conclusions have been based upon the presumption of the absolute indestructibility of mechanical energy;—or the impossibility of the destruction of one form of mechanical motion, unless at the instant of its disappearance there be generated an equivalent motion of another form, into which the first is converted. This is the modern doctrine of the conservation of force, than which nothing is more fundamental in the scientific mind. In accepting it, and following to its consequences the nebular hypothesis, and the accepted dogma of conservation, we arrived at the universal equilibration of matter;—a condition of eternal monotony and nullity, requiring for its re-vivification, the aid of the Omnipotent. And to the further conclusion, that unless the grand law of gravitation changes in each almost incommensurable cycle, from concentration to dispersion, direct agency and interference are, in like manner, indispensable. Thus, when the processes of nature are measured by almost inestimable spans of time, we find the stability of what has seemed eternal constellations, vanishing like baseless fabrics. The majestic force of these mighty orbs, transformed into puny vibrations of atoms, and themselves dissolved into vapor; or else, commingled and concentrated into one immeasurable mass, without light or life, pursuing through space an objectless, sullen way.

We find the energy of matter, under the great law of gravitation, unable to perpetuate the life and vitality of nature. And no one will for an instant suppose, that matter could of itself divest itself of one law of action, and assume another. We are driven to the alternative and conclusion, then, that, unless the great laws of the conservation of energy, and of gravitation, are misty nothings, Divine agency must, from period to period, move within the infinite depths to resuscitate matter to life and purpose. Thus under the laws of science, considering matter as a whole, a vast aggregate, we are led by her ways, to universal spirit, without whose direction, matter and its laws are naught.   And upon the dizzy heights to which we have been conducted, we have the most sublime conception, overshadowing all others, of a Supreme Ruler of the universe.

CHAPTER III.

FORCE.

MOTION appears as a conditioned manifestation of force, seen through matter, and in
terms of time and space. All conceptions
of matter are from resistances of positions in space,
contrasted with positions offering no resistance.
These positions of resistance have extension, and
are seen to have no independent volition. The invariable passiveness of this extension in three directions, to which the term "matter" is applied, and
the energy required for its motion through space,
have created the distinction between inertness and
force. To the passive helplessness of matter there
is given the term "inertia." This is not a force, but
the absence of force, though it measures the force exerted to produce motion. Experiences of force presume, that substance at rest, is without the ability,
of itself, to move ; and that substance in motion, is
without ability either to arrest or to vary its motion.

35

Observation having established the inertness of matter, any mutual influence upon each other of discon-
nected masses or atoms, is necessarily imputed to
the presence therein of an energy.   Since all of our
experiences of force are derived from matter, all
physical forces not proceeding from matter, are,
strictly speaking, inconceivable.   And as such forces
cannot exist without matter, so if matter were not
inert, or had no inertia, force would be useless, as it
would have no function.

Nothing can be known of the final nature of force ;
it is an attribute fixed in matter, yet not matter.
Whatever its nature, it can no more change the direction of its influence, or regulate its own action,
than a final unit of matter can contract or expand
of its own volition. All motion is material, and represents material energy.   Substance cannot, therefore,
part with its motion without an equivalent energy
being imparted to other substance. " If those
motions, through which the parts pass into a new
arrangement, might either proceed from nothing or
lapse into nothing, there would be an end to scientific interpretation of them. Each constituent change
might, as well as not, be supposed to begin and end
of itself.   That the relative reality which we call
motion, can never come into existence or cease to
exist, is a truth involved in the very nature of consciousness."   Throughout nature there is no example of absolute rest, all asserted rest being merely
the expressions of relations of bodies to other parts
of space. Atomical motion attends all thermal variation ; this variation alone is incessant and universal·

Chemical and polar motion are unceasing; and the diurnal and annular motion of the earth perpetually change the position of every atom of its mass.

The interconnected movements of the solar system, and the motion of that system toward a distant constellation, together with the motion of stars and nebulæ, are evidences of continual transition, from which we reasonably infer a motion of the whole stellar world, the verification of which is prevented by the absence of appreciable parallax, and by the limited period of our observation. The universe is relieved from monotonous sameness, and is endowed with activity, whirling life, and beauty, simply by virtue of the never-ending motion of each and every unit of matter.

Early observation, though not less satisfied of the motion of matter, failed in its interpretation.

The familiar phenomena of change in the four supposed elements of earth, air, fire, and water, as well as the periodical motions of his planet, affecting his comfort, forced upon man conceptions of motion, as old as his race. Heavy bodies, unsupported, were seen to fall; hence it was presumed that all heavy bodies tended downward and conversely, that if bodies tended upward, it was because they were light. "So satisfactory and self-evident was deemed the inference of weight, that, as has been remarked, of all gods, a god of weight was never originated." If a body in motion stopped, it was concluded that its force of impulsion had been not only exhausted, but annihilated. Its retardation was looked upon as a disposition to rest, inherent in the body. Motion was long considered an effect, definitely beginning and ending; and of course the possibility of continuous motion without the continuous operation of the cause, even though the body was undisturbed, could not be entertained, since a resistance equal to the lost motion was unthought of. This reasoning extended to all the changes of position noticed in celestial bodies; and substituting occult qualities for inherent energies, attention was occupied in baseless hypothesis, encumbering observation, and retarding the acquisition of exact knowledge.

We omit, as out of place, tedious and complicated, the various and ingenious theories advanced to account for the motions of the solar system; and the steps in discovery, which led to the laws of planetary motion. Through the genius of Copernicus, of Galileo, and of Kepler, all astronomical questions of movement were finally merged in those of rational mechanics, in which motion is inseparable from the force producing, or tending to produce it. But in the mind of Newton, concurrent with the observed gravitation of planets toward the sun, was raised the question of the cause of such disposition. He believed the cause to reside within the bodies themselves, and reflecting upon the falling apple, he believed gravity to be identical with the planetary propensities. He demonstrated the truth of his supposition with respect to the moon, and then to the planets, that a force having the same law of variation as to quantity and intensity, governs the motions of all. And he justly concluded that the force of gravitation within our solar system is an identical force, as like effects proceed from like causes. From this great explanation of the cause of gravitation of bodies, the idea, for the first time within the history of progress, was finally seized upon, that matter was the seat of force ; and here was the idea of force, as a cause of motion, pursued to its logical sequence. Here the ultimate datum of consciousness is expressed in the analysis of a daily observed fact. It is a declaration to all time, that matter is an inert existence, in which force resides, causing its activity. The great law of gravitation discovered by Newton, is a law of the attraction between all masses of matter. The force, or attractive action, varies directly in the ratio of the masses acted upon ; and it varies inversely as the square of the distance through which it acts; thus, if of two masses, the amount of substance of both of which is represented by ten, the amount of their substance be made twenty, the attractive effort between them will be twice as much ; and if the distance between any two masses be at first ten, and it be diminished to one, the amount of attractive effort will be one hundred times greater than before. And it is also a consequence of this law, that right lines between masses are lines of intensity of attractive force. The universality of attraction of gravitation is now conceded in scientific thought. Proof from many sources is multiplied to establish the truth of this most general physical law known to man. The deflection of the plumb line by masses of mountains in its vicinity, shows its earth action. That it is a law of sidereal matter, appears from observations upon double stars, their revolutions, and orbital cyles; from cluster stars, their tendency to condensation toward a nucleus, and their globular figure; from the spiral form assumed by nebulous matter; and from the spiral figure of an extensive class of galaxies of stars; from the observed common tendency of many stars to recede from, and of many others to advance toward, a certain point in the celestial sphere; from the motion of our solar system; from the motions of many stars hitherto fixed, so far as our observation extended. Analogy indicates a harmonious similitude between the laws of our system, and those governing, in remote regions, the same matter as that of our own system. "The universe is a sphere, whose centre is everywhere, and whose circumference is nowhere." Diversity in laws of motion and force in separate systems, all in motion, could not long sustain order in the universe; sooner or later would ensue confusion and destruction. Everything indicates a grand and simple unity. Regularity and harmony can be universal, only on the ground of one controlling principle.

It will be observed of the force of attraction of gravitation, that its exertion through space is independent of a material medium. It is not at all contingent upon atomic transfer. This, of course, is not comprehensible, but we are compelled to admit this truth. It is an immensity of energy, operating between vast aggregations of matter; and no attenuated medium, as interstellar ether, could convey, or contain forces so disproportionate to ether substance. But supposing that there was a dense intermedium between these great aggregations of matter, would the transfer of this attractive force through such a medium be any more conceivable than before? One of the two suppositions must be adopted, yet one is no more intelligible or conceivable than the other. Both imply an influence of traction extended through space, and whether supposed through a material intermedium or not, it is made no more intelligible.

The common experiment of the magnet and a piece of iron demonstrates the full possibility of such action without a continuous intermedium.

All matter, then, is constrained to follow the direction of action of an inbeing force. Unless we assert, that that which is moved, and that which causes movement are identical, which is contrary to experience. The one expresses passiveness : the other, the capacity to impart motion and to govern its direction. Force implies a resistance ; and the operation of force, an effort opposed to a resistance. The evidence of the senses is that the universe is made up of matter and force ; or of force and that which is in itself without force, the last the burden, the first the carrier. To affirm that matter actively resists, or exerts effort, is to affirm that it is a force. If it be a force, the resistance of a mass to movement must be accepted as the result of a multitude of merely mathematical points of resistance, aggregated, yet without substance. Each point under this supposition is an effort of repulsion, opposing efforts of impulsion upon it. And the world of matter consequently presents nothing more than immaterial points of force, opposing themselves, without any apparent cause, to other points of active effort, similar to them; sometimes overpowering them, and sometimes failing to do so; sometimes visible, sometimes not so,—all obvious absurdities. The only alternative then is to accept, as existing, substance, real, final and inert; whose resistance to motion is merely passive, and conveying no idea of real energy. Force and matter compose the whole of the phenomena of nature, and indeed of the natural universe, so far as it physically presents itself, in all its variety and beauty. And if the phenomena of nature be subject to invariable law, then the forces of matter are definite in quantity and measure, and invariable in action; otherwise order and regularity must proceed from disorder and a fantastic uncertainty. Admitting that there is an attractive force exerted between masses of all matter, it follows that this force is a resultant of the attraction of all the component forces of the final units of the mass; for all masses are made up of such units, and in them must reside the force of attraction; in other words, they are the real loci of the attractive energy: and any one of them exerts an energy in the exact proportion to its quantity of matter, or magnitude. This energy of simple attraction residing in the final units of all substance, will be hereinafter designated, for the sake of brevity, as *primary force*, in contradistinction to *secondary* force, represented by motion of matter, which it may modify.

This primary force, or principle of energy must

always be resident in matter;—if not, it may exist outside of matter, as an unlocalized entity, or existence;—an agent of pure immateriality, and of course, in all its relations, necessarily indifferent to law. Matter, then, would be sometimes influenced by it, and sometimes not. It would be sometimes deprived of motion, and sometimes tend to excessive motion, under its influence; this is contrary not only to all law, but to all experience,—and therefore such assumption must be rejected. Similarly it may be shown, that the resident primary force of any ultimate unit of matter is always the same; whatever its power is, *that* it must forever remain. Otherwise matter may of itself augment or diminish, possess or dispossess itself, *ad libitum*, of energy. As the primary force is then for each unit, a definite power, and the same for each mass, the quantity of primary force in the universe is unchangeable. Nothing can be clearer than that there can be no increase, or decrease, of this innate force of matter. If so, matter may have originated, and may still originate motion; which is contrary to any idea of harmony or law. Of course, we do not here refer to the mere intensity of attraction between masses or units, which is greater as the interval between them is less. And here again, in this very decrease of attractive energy, or primary force, with increase of distance, may be recognized all the peculiarities of a limited force ;—a diminished capacity to act.

Let there be now considered this primary attractive energy, whose presence is confirmed in all matter, bearing in mind, in the exploration of this unknown, that the uttermost limit of research and argument can attain to nothing more than reasonable probability ; and that this is also the limit of all science not susceptible of experimental demonstration. Let there be imagined a final spherical unit of matter at rest, and isolated from all influences, and though its moving principle be absolutely immaterial, yet for illustration, likening it to a viscous elastic fluid, it is obvious that this principle of energy will be symmetrically distributed with respect to the unit substance. To suppose otherwise, would be to assume an effect without a cause. Besides, if the resident energy were unsymmetrically distributed, the unit would assume motion in the direction of most energy. Now if the second like unit be placed within the influence of the first, the force of each becomes more concentrated in the segment of each nearest the other, and motion ensues, because of the preponderance of force, in the direction of concentration.

Since matter can, of itself, have no power to produce motion, the cause of motion is the sympathetic exertion of *primary force, towards primary force* in other matter. Not of *matter* towards *matter*, for matter is inertness ; *not of matter* towards *primary force*, for the same reason. The tendency to the equality of distribution of force in matter, *is a condition of the existence of associated force and matter ;* and *any law* of the action of force, *however fundamental, is subordinate to this condition of the fixation of primary attractive force in matter ;* otherwise, in the tendency of force toward force, it would desert

its own matter. In the tendency of force toward
other force, there is *always a limitation of its action*,
which is the degree to which the symmetry of distribution of force in each unit of matter may be
disturbed. The universal inclination of matter to
union, would be consummated by the complete contact of the final units of matter, and the end of all
life, were this not prevented,—as we hope to show—
by this principle of limitation. Though the *perpetual tendency* of the primary attractive force of every
unit, must be to symmetrically distribute itself with
respect to the unit magnitude ;—the consummation
of this effort is as perpetually prevented by the in-
terassociation and motion of all matter.

Let us now return to the supposed case of two
units of matter, isolated from all other attractive influences. As they approach each other, the energy
in the part of each situated nearest the other, is
greater than in the opposite part. An energy due
to the closer proximity of these parts ; an intensity
due to their less distance. The closer the units approximate, the greater this intensity ; when almost,
or quite in contact, their forces have become almost,
or quite, one force, with by far its greatest intensity
near the almost, or quite, joined segments. This is
contrary to the law of existence of force and matter ; that is, that force *shall be symmetrically distributed with respect to it :* hence, when the degree of
approximation of the units is to actual contact, the
forces will no longer strive to unite in the middle of
the units, but will exert themselves to equal distribution in the two units. Consequently as these ener-

gies tend to recede to more uniformly distribute
themselves, the units themselves will, under their influence, recede from each other; and this motion
will continue, until again counterbalanced by the integral attractive energies of each unit, and these last
positions will be their positions of equilibrium with
respect to each other. But it is self-evident, that
these positions of equilibrium could never become
positions of rest; and that if two units could be
isolated, as these are supposed to be, that they
would oscillate back and forth, to and from each
other, forever.

It is impossible, but that, as the two units approach to contact, there should be a greater intensity of energy at the parts of the two units nearest
contact; and that there will be a point when the
energy of each unit mass will resist this inequality
of intensity, since it is opposed to the equality of
distribution of energy; and that they will, therefore,
resist a motion which tends to make greater this
inequality. For as the matter of both units tend to
become one unit mass, to that degree do their energies tend to become more the character of a single
unit energy, and will resist, accordingly, what is
contrary to *the condition of existence* of that energy,
that is, *inequality of distribution.* It might appear
that the position of equilibrium with respect to
motion, of these two units, would be when their
surfaces were just in contact. This would be the
case, were it not for the fact, that intensity of action
of the force of each unit is greatest on the segments
in contact; therefore the quantity of force there is

greatest, and this must cause the recession of the units to some point removed from actual contact. This causes the never-ending oscillation of all final units of matter, without which neither life in nature, nor difference of density of substance, would be possible.

We now turn to inquire whether there could exist in matter an absolute primary repulsive force, as marked, as distinct, as sharply defined, as is primary attractive force. For it is asserted by physicists, that when units of matter approach to final contact, that instantly (and of course without explainable cause) the admitted attractive power, innate in each unit, is at once paralyzed, and that there is instantly generated (from nothing) the unlimited repulsive force, compelling the units to separate; and that then, in turn, the repulsive force becomes dormant, and the attractive force is awakened to effort. All this is without cause, inconceivable, and entirely presumptive. But it is a part of the physical philosophy taught at the present day, which assigns to each unit of matter, not only the distinct and separate forces of attraction and repulsion, but also *special magnetic*, and *special electric forces;* and a *special chemical force*, or affinity, peculiar to each kind of matter; all of these distinct and separate special energies, co-resident in each unit of matter. Is it any wonder that with the scientific imagination running riot in such absurd assumptions, that ignorant and blatant scoffers loudly declare the omnipotence of matter, and denounce the necessity for either spiritual wisdom or power? Faraday, pro-

foundly convinced that notions of an attractive
force, suddenly disappearing upon the near approach
of units of matter, ignored entirely the principle of
the conservation of force, says : " The idea of a force
simply removed or suspended, without a transferred
exertion in some other direction, appears to me to
be absolutely impossible." The idea of a force,
whatever its nature, changing its direction of action
from attraction to repulsion, passing through zero to
an opposite energy, involves a change of the nature
of force, which is contradictory to permanent law ;
and it is wholly antagonistic to the law of conservation of force, for there is no " transferred motion."
In the case, however, of our supposed two units of
matter, it will be observed that there is a " transfer
of exertion " of their forces on contact of the units,
and that is, toward equality of distribution of force
with respect to matter.   It is plain that if there be a
distinct force characterizing each different elementary mass of matter, (a peculiar form of force for
each element), then each of these peculiar forces
must reside in the units of the mass ; for in all cases,
peculiarities of qualities must be referable to the
composing units.   Therefore, under existing views of
matter, each final unit must be invested with four
distinct energies : attraction, repulsion, chemical, and
polar.   The harmonious association, yet independent
action of these forces, each of which is, at times,
predominant to the exclusion of the others, not only
transcends analysis, but opposes rationality.   Yet if
we accept a duality, as primary attraction and repulsion, we may as well accept any plurality.   One is

quite as conceivable as the other. Polar and chemical forces unquestionably exist ; but we shall endeavor to show that they are contained within the law of action of one primary attractive force. Of course, it can never be positively determined that a variety of inherent energies does not exist in each final unit of matter. Actual analysis can obviously take us no further than elemental matter, and this we can examine only in a state of aggregation ; and only through, and by means of those very forms of energy, heat, light, electricity, magnetism, and chemical affinity, all mutually convertible, and which have been, and are still received, as exponents of the inbeing qualities in the units, and of the same nature as their aggregations. Color, hardness, taste, fusibility, etc., are not to be considered, as they are qualities of aggregate matter, which vanish with its disintegration. But there is a negation of a multiplicity of primary energies, to be found in its very irrationality. This doctrine has arisen as a convenience of hypothesis, and now having served its purpose in the growth of science, it should be permitted to pass away, for it is no longer justifiably maintained. It elevates the purely material to the region of the vague, the indeterminate ; and presenting at the outset an incomprehensible barrier to any understanding of either matter or force, merges both into the mysterious, and, with the ignorant, even into the supernatural. The evidence of a primary force or gravitating force of matter, has a broader derivation, and a completeness that belong to no other. It is unchangeable in character, belongs to all matter

4

equally, and is independent of any state of change,
or any combination. This is not true of any other
asserted primary force whatever. The evidence of
their condition is only partial. They are sometimes
manifested and sometimes not, being always conditioned by modes of motion. The very fact that all
the forms of energy of heat, light, electricity, magnetism, and chemical force, are mutually convertible,
alone, almost demonstrates that each of these forms
is due to a *mode of motion* only, of matter, which is
under certain favorable circumstances developed ;
and that the phenomenon is not in itself due to a
peculiar inherent energy.

The primary force of attraction of gravitation is,
on the contrary, in no way affected by the motion
of matter, nor can it be in any manner either produced or suspended by any form of motion, nor can
it be converted into any other form. It is especially important in everything pertaining to physical
knowledge, that the most fundamental ideas should
not be held to be quite as likely delusions of immateriality, as the probabilities of absolute reality. Since
it is from physical knowledge that we rise to a higher
and more abstract order of thought ; and if it be
unsettled, or received as unreal, all other fields of
study are the more hedged about with doubt, or are
the more open to the growth of scepticism and
credulity. Physical learning should be a recognized
actuality,—defined, clear, and limited ; and the public instinct should never be vitiated, by investing
physical phenomena with mystery and wonder, by
delivering ourselves to explanations more strange

than original difficulties, as in the imaginings of
Berkleyism, or the theories of Boscovich of immaterial forces. But two general methods of procedure
can be progressive ;—one, to unreservedly aban-
don all theoretical views not susceptible of complete
demonstration, and the other, to adopt those reasonable and probable, and having a basis, defined
and actual, in the elements of matter, force, and
motion. The last is the natural method, and corresponds to man's apprehensions of a tangible and
material environment, and to his relations to it;
and at the same time it furnishes ample scope and
stimulus to his imagination, that powerful aid to
discovery.

We now pass to the consideration of other points
relating to primary attractive force ; the first is, its
aggregate action. If we conceive two final units of
matter, the one very small and the other large,—of
even a sensible magnitude, the inequality of their
attraction at any distance will be due to the quantity of force of each, this quantity, according to the
law of the force, varying with the mass of any
volume of matter. Of course the attractive energy
of each is definite ; to assume otherwise, would be
to assume that it extends to an infinite distance.
This is contrary to the law of its diminution, which
is, that it varies inversely with the square of the
distance. That is, if at the distance designated as
one, the attraction between these units be ten, at
the distance ten the attraction will be one one-
hundredth of what it was before. So then it follows, that at some distance, where the attraction

of the smaller unit would be absolutely nothing, that of the larger is a real amount. What is true of the forces of these single units, is true of masses; for, as the units join, there is a confluent action of their forces; and the range of attractive effort is therefore extended by the increase of mass, just in the ratio of that increase; for plainly, such increase is the *increase of the quantity* of primary force at the origin. The capacity of the attraction of a mass then, comes from the association of the unit energies, and is a resultant effect of these components. The variation of intensity of the force with distance, may be compared with that of a definite amount of light; it may be diffused at the expense of intensity, or concentrated at the expense of quantity. When the distance between two masses of matter is lessened, and the intensity of their attraction becomes greater, it is not that there has been any creation of force or energy, but because that degree of intensity, or of energy, is an energy *due to that distance*, and is *a potential force of matter*, always the same, whether masses are within each other's attractive range or not. There is great reason for believing, (but the evidence is too voluminous to be detailed here), that matter, instead of being diffused into its final units, and those units acting separately, or as so many unit energies, is in a molecular form, in its minute activities, or in the form of a vast number of units united into one atom; and that in these molecules, or multiples, the units of each act in concert in all the *different affections* of matter, such as light, heat, chemical, magnetic, and electric manifestations. This hypothesis is well sustained and is contrary neither to observation nor to experiment. A volume of water dashes from a height, encounters resistance from the air, breaks up into masses, and then into drops. Increase the dispersive force, and the drops disintegrate into a lesser magnitude, correspondent to the motion; and by a still greater agitation, as when water forms steam, and steam is heated to a high degree, the particles become lost to perception. The self-maintaining power of a volume being relatively greater as a body is smaller, the effort of the attraction soon disappears in the spherical figure of the drop. And none can presume that that figure is lost in any succeeding division of the unit drops. However far the agitation extend, it is still a multiple spherical atom so long as its hydrogen and oxygen are united.

Mercury, less liable to "wet" surfaces, sustains itself in larger spherical globules, which are smaller in the ratio of greater motion to which they are subjected. By great heat (vibration), Mercury is volatilized and invisible, but the form doubtless persists as spheres. The solution of bodies in liquids, the freezing of water, the isomerism of bodies, and many other phenomena confirm the idea of the multiple atom or molecular constitution of matter; and as the forces of each unit are central, it follows that the form of multiple atoms is the sphere. Of course, it will be understood, from what has been said in the foregoing pages, that the final units making up any multiple atom, or molecule, though acting in each molecule as a unit of force, are by no means

joined together by actual contact. Intervals of
thousands of times their diameters may not express
the distances apart, of these units, through which
their individual power is so intense. And it will
be remembered further, that not only are the units
of each molecule constantly in a state of inconceivably rapid vibration, but that the molecules, or
multiple atoms, themselves, of any mass, must also
maintain a state of swift oscillations;—for the reason, that no two multiple atoms of any mass in the
universe could, by any possibility, have precisely the
same amount of motion; since no two portions of
any mass of matter, however large or small, would
receive from surrounding media precisely the same
amount of material impulses, or incident motion;—
and that the motion of the molecules and units is
brought about by their constant interchange of difference of motion in the mass, and its tendency to
equilibration.

It is now necessary to call to mind what has been
said of the intense heat consequent upon the impact
of sidereal bodies, and the immense diffusion of
matter accompanying this heat; bearing in mind
that what we call heat, is nothing more than vibratory motion of molecules. In the cooling of this
matter just the inverse process takes place to that of
heating it, or as in the breaking up of water and
mercury, as has been described,—that is, the final
units of matter would first coalesce; then molecules;
then groups of molecules having about the same motion; and last, great aggregates of these groups would
form into masses. To explain this further. In the

collision of celestial bodies above referred to, their
motion of translation is transferred into the shiver-
ing, vibratory or heat motion; and this heat is estimated to be more than twenty thousand times
greater than could be produced by artificial means.
Of course, all endeavors to ascertain the relative
density of the incandescent matter when in this
condition of extreme heat, would surpass conjecture
itself.

The intervals between the final units of the most
solid masses of matter have been compared by
some physicists to the spaces between stars in the
heavens;—and of course, in the utter disintegration
of stellar matter following collision, these intervals
are greatly increased. But however great the heat,
and diffusion of matter, the attraction of units for
each other will maintain its continuity; for without
their mutual relation as to motion, there could be no
extension of volume. With the partial destruction
of heat motion, by radiation into space, beyond the
diffused heated matter, begins, as has been said, the
first stage of contraction. And as the forces acting
are but two, that of mechanical heat motion obstruct-
ing, and that of attraction promoting the concentration, it follows from well known laws of mechanics,
that these beginnings of aggregation will be the separation of unlike units from each other, or of units
of unequal magnitudes, and the formation into
groups, of like units; and this process will continue,
throughout the entire nebulous volume however
large. For similar units will be similarly acted upon
and will similarly act and react; and under the laws

of force, the differences of motion of different orders
of units as to magnitude, must result in their formation into groups of units of the same order. For
similar reasons groups thus formed of the same order
of units will contain approximately the same number
of units. Groups of the largest units will be of the
largest dimensions, and these groups and their units
will have the least motion; while those composed
of units of the smallest magnitude will be of the
smallest dimensions, and have the most motion.
These primary groups of matter, the first stage of
concentration, we designate as molecules, or *multiple
atoms.*

It is plain that they are not again resolvable into
their final units, save by a return of the great heat
of the diffusion in which they were once immerged.
The multiple atom of matter, then, *is indestructible,*
so far as we are concerned, and it is the unit to which
we must look in all considerations of primary force.

Following mechanical laws of force and motion,
these primary groups of multiple atoms of like units
will finally unite, as in course of time the heat of the
nebulous matter diminishes, and they will form
elemental matter of like multiple atoms. By great
heat then, all matter can be volatilized ; and by the
greatest artificial cold, all gases can be reduced to a
liquid state; and by analogy, if we had the means
to bring about an absolute zero of temperature, we
could solidify all matter. We must conclude then,
*that the present state of aggregation of matter is
due to a certain amount of heat retained in, or to
oscillatory motion of its multiple atoms;* and as no

substance has absolute hardness, there is vibration of the final units and multiple atoms of *all* matter. There remains one other question relating to the forces of a multiple atom, or molecule, so material to an understanding of the action of forces in chemical, and other phenomena, that at the risk of being tedious, we find it important to our purpose to explain. For while such matters may be intelligible enough to the man trained to science, they are not so to the average reader. And the whole idea here is to appeal to that class, who, with general information, are inclined to general scepticism; and who yield to science so many possibilities, that when they hear it proclaimed by the atheist that nature is an all-sufficient power, they are very apt to entertain to some extent his doctrine. Let us return to two units of matter conceived to be isolated in space from all attraction, except that of each for the other. If the unit *a* is of greater magnitude than the unit *b*, it exerts toward *b* no more attraction than *b* extends to it. Sympathetic tractive effort cannot excite a greater response than it conveys; otherwise there is no reason why the earth should not move with as great rapidity toward a falling object, however small, as the object toward the earth. Let it now be supposed that one of these units, as *a*, is the central unit of a compact mass or sphere of units. Plainly, since the whole energy possible to any final unit of matter is a fixed and definite quantity, it results that the sum of the attractive effort of all the other units for this central unit is exactly equal to its energy, and no more :—and that its energy being so

divided is for each of them only a fractional part of
what it was for the unit *b*, in the first supposition.
If now we suppose a spherical shell of units one foot
in diameter, exactly conterminous with the sphere of
units, yet still within the attractive limit of the unit
a, its whole attraction is again divided between the
units of the shell, as well as the sphere, and is less
for each than before, for otherwise there can be no
limit to the power of any unit.   It must further be
obvious, that when we speak of the *intensity* of attraction at any distance between two masses or units of
matter, we can mean nothing more than the *quantity of attractive effort at that distance*.   It follows
therefore, since the whole quantity of attraction possible to any unit is definite and fixed, like that of
any other force, that if a part of this attraction is
employed in any one way, it is tied up to that extent
from exertion in any other way :—and that just to
the extent that it is released from such employment,
it has that much more energy free.   Now let it be
supposed that the small sphere or mass of units
around the unit *a*, expands to the limit of the spherical shell ; plainly with the increase of distance from
a, of the units, some of the attraction of *a* is released from employment and restored to itself.
What is true of the unit *a*, is true of all the other
units of the small sphere ; each has restored to it by
the expansion of the mass, and separation of the
units, a portion of its energy, which is withdrawn
from attractive effort, by reason of the greater dispersion of those units.

   From the foregoing it will be seen that in a multiple atom, or molecule, composed of many final units, in the ratio that those units are separated by expansion of the multiple atom, due to an increased vibration, to that degree are their energies released; —released from concentrative employment, and may therefore in each unit act more individually. The consequence of this would be that the whole energy of this multiple atom, toward another situated within the limit of attraction of its final units, is greater than before the expansion of the atom; since each unit has more free or unemployed primary or attractive force than before. We have been thus diffuse upon this subject at the risk of wearying, because it is upon this principle, as we shall show, that depends the chemical energy of matter; and not because the final units of each elemental substance are invested with a peculiar inherent force distinct and separate from any other substance. As we were conducted to a view of the Infinite Spirit in reasoning upon the majestic forms of matter in its vast aggregation; so in these final and inconceivably minute forms, it may be shown that there is the perfume of spirit power, though masked for a time, it may be, under the domino of attractive energy.

CHAPTER IV.

THE FORCES OF THE PHYSICAL AFFECTIONS
OF MATTER.

Heat and Light Motion—Interchange—Manifestation of Electrical
Phenomena — Mode of Motion—Proceeds from One Primary
Force—How Produced—Mutual Interchange of all Physical
Forces of Matter—Probable Nature of Magnetism and Elec-
tricity—How they Differ—Direction of Transfer of their Ener-
gies.

THE physical affections of matter, now known as
modes of motion, are light, heat, electricity,
magnetism, and chemical interchange. Our
only purpose in briefly alluding to each, is to show
that matter and force are things of the utmost simplicity, so far as they are presented to our senses by
their phenomena ; that there are no occult properties,
nor mysterious fluids, nor indescribable entities,
whose power and agencies are immeasurable and unfathomable. Of course, we know nothing, and can
know nothing, of the final nature of either matter,
or its inherent primary force ; but we do know, that
both matter and force are governed by laws ; that
their creation or annihilation is unthinkable; that
matter is real, so far as anything can be real to us,
and that each of its final divisions, or ultimate units,

is invested with an inherent and definite energy, that is forever fixed in quantity, and that it can never be increased, or diminished, under existing laws of nature.

We purpose to show how, under the operation of one innate primary force, attraction, all the above-named affections are produced; that each is a mere mechanical effect of the motion of matter and this one force; and that as neither, under any circumstances, can produce, nor bring about, life, there is no reason to believe that operating jointly they would of themselves approximate any closer to that result.

HEAT AND LIGHT MOTION.

Heat and light are merely modes of vibratory motion of matter, though the precise nature of this motion is beyond analysis, the inquiry being limited from the very constitution of things. The communication of heat is the communication of motion, and its effect is to separate from each other the multiple atoms, or constituent molecules of matter. It has already been shown why, upon the contact of final units of matter, they repel each other; and for the same reason, two multiple atoms, each containing a vast number of such units, would, when thrust together, repel each other. This makes the elastic effect of matter. Vibration without elasticity, or the power to return, is impossible. When the multiple atoms of any volume receive an impulse, from other atoms communicating more motion to them, their vibratory swings are longer, and more energetic, they occupy more space, and the whole volume of the matter affected, expands. The effect of heat upon all matter, with one or two exceptions, is to expand it ; and casting aside all ideas of heat and cold as sensations, heat is simply the motion of expansion of bodies, through incessant atomical motion. Justly then, all inquiries upon heat relate solely to degrees of motion. Heat motion applied at any point of a solid volume of matter, thrusts the atoms there further apart ; opposed to their motion, is the united reactionary effect of the other atoms, or their tendency to maintain their positions against displacement. But if the extraneous motion be continued, it overcomes the successive degrees of resistance, and disintegration follows.

With yet further persistence of the motion, the vibration of the atoms is increased, each occupies more space, and the volume further rarefied, passes from the liquid to the gaseous state. Contrariwise: as heat signifies motion of atoms, so cold implies the absence of such motion ; and other things equal, the lower the temperature of any body, the nearer must be their constituent atoms. Since heat and cold represent degrees of motion, no real distinction exists between degrees of heat and degrees of cold ; and any instrument for their measurement, estimates simply the relative amount of motion in the matter tested. What has been said of heat and cold being merely relative degrees of motion, applies, *mutatis mutandis*, to heat and light ; they are merely degrees of motion, since most forms of matter become incandescent by increasing their amount of heat motion.

When a body or surface reflects vibrations of certain degrees of rapidity, the eye is affected, and we see the surface presented to view, color being, of course, modifications of this vibration.

Consciousness of heat, of cold, of light, of color, then, is only the consciousness of amounts of impinging motion, from material media surrounding us. Just as there is an upper and a lower limit to the range of hearing, and in light, degrees of intensity in both directions, beyond our perception, so in heat there is a maximum and a minimum of intensity impossible for us to realize, the limit of experience of heat motion being but a finger's span to the inestimable range above and below it. Even if it be supposed that all ultimate units of matter are of the same magnitude, it would still be impossible that any two aggregations of matter in nature should possess the same degree of heat motion, for the reason that they could not simultaneously occupy the same position in space, and would therefore be subject to unlike incident forces each instant of time. Still less is this possible under all the diversity of the differences of magnitude of the final units of different elements of matter.

This inequality of heat is, however, regulated, and meets with speedy distribution by the never-ending process of interchange of motion through the interconnection of all matter. Each body possessed of an excess of heat motion tends to impart it to contiguous bodies of less motion; not because these bodies extract, or withdraw of themselves its motion, but because motion always follows lines of least

resistance; and the matter presenting least resistance to the heat and motion is,—other things equal,
—that having the least motion of its own to counter-
act, before its motion becomes synchronous with
the imposed motion.   The direction of interchange
of heat motion, then, will always be from the mass
or volume of *greater* heat motion, to that of *less*
heat motion.   As the units of different elements of
substance are, under our theory of matter, of different magnitudes, the units of each element have
different degrees of vibration, even though each has
received the same amount of heat motion.

For the same force will impress upon different
magnitudes, different velocities; yet the product of
each magnitude by its velocity, will be the same as
the product of any other magnitude by its velocity.
Therefore, under the same temperature, the state of
aggregation of different elements is dissimilar; some
must take the solid, some the liquid, and some the
gaseous form.   It will probably have been concluded
before now by the reader, that such a thing as the
isolation of any unit, or mass of matter in the universe, from any other unit, or mass, is an impossibility: each unit is, through its primary attractive
force and motion, constantly in touch, with respect
to influence with other units.   Refined and conclusive experiments indicate that all space is occupied
by a substance (ether) surpassingly rarefied and elastic, through which the radiant impulses of light and
heat make their way.   By its elasticity, this medium
is uniformly diffused, both in stellar space, in what
we call vacuo, and in the interstitial space of the

multiple atoms of all bodies, whether solid or fluid. Heat and light motion are, by it, transferred not only through our most perfect vacuum, but they traverse in all directions inter-stellar space; it is a universal bond of dynamic force. All evidence culminates to render it conclusive, that light and heat are only different degrees of the same form of motion or dynamic energy. The laws of their phenomena are the same in all respects.

And there are no data from which it can be supposed that light may be generated by other modes, or in other ways, than those developing heat ; and there is no change or manifestation impressed upon light motion, that does not effect the associated heat in the same manner. The incident velocity of both is the same ; both are diffused at the same rate in the ethereal medium ; and both are by the same laws transmitted, reflected, absorbed and refracted.

ELECTRICAL AND MAGNETIC PHENOMENA, AND THEIR EXPLANATION UNDER ONE PRIMARY FORCE.

In considering electricity and magnetism, no more than in the other physical affections of matter, do we propose to discuss the forms and laws of their manifestations, and the many variations of these laws, and the vast array of facts which unite to confirm the view of a single primary innate energy in matter ; but simply to point out the correspondence of their phenomena to that of matter directed by a single primary attractive energy.

There are probably in nature no physical manifestations, which have so axiomatically fixed images in the mind of distinct repulsive, and attractive energies, as those witnessed in magnetism and electricity. Though it is true, that some advance has been made, in abandoning the old notion of incorporated fluids ; and it has been well said, that, " The supposition of a gravitating fluid might, with as much propriety, be insisted upon to explain gravitation, or a cohesive fluid to explain cohesion." But there are great numbers of otherwise well-informed people, who entertain the idea that there actually exist entities like electricity and magnetism, chemical affinity, vital force, etc., attached to matter, and yet not of themselves material, though subject to material laws. To these various entities, prodigies of performance are conceded by such minds, which are satisfied to extend their inquiries no further.

Electricity is now defined in science as being a " compound force, remarkable for the peculiar form of action and reaction which it exhibits ; this kind of action and reaction follows the same law of equality and opposition in its manifestations, as that which is exhibited more obviously in the phenomena of mechanics." This explanation implies two opposite forces, attraction and repulsion. That there may be these two forces located in *all* matter, (for *all* may be to some extent both magnetic and electric) one of two conditions of matter must exist; either all masses must have just as many units invested with inherent repulsion as there are units invested with inherent attraction ; or, each unit must be equally invested with an inherent force of repulsion, and an inherent force of attraction, co-equal and co-resident.

Either of these conclusions is utterly illogical and unthinkable. Besides, if matter were so conditioned, there could be no resultant attractive energy in the mass whatever. Without further discussion, we proceed to illustrate the phenomena, under the presumption that all matter is identical; and that each final unit is invested with one force of attraction varying in quantity with its mass or magnitude. If any two molecules, or multiple atoms of like units and motion be within the range of each other's attraction, they will continue to attract until contact. The aggregate force of each atom acting toward the other, being the sum of the unit forces *free to act*, (see last part of chapter on force) or the *resultant* free force of each. As the two atoms come into contact, the two resultant primary energies tend to distribute themselves so as to establish the uniformity of force, with respect to matter.

But since in their nearest segments, are the greatest quantity and intensity of energy, and as neither atom can merge into the other, (they being indestructible,) the position of equilibrium of these two atoms is that at some point of actual separation. And they would therefore recede from each other immediately after contact; as already explained in the case of two final units of matter, in the chapter on force. The atoms are supposed, for simplicity, to be of like units and motion. If, however, the units of one atom, as *a*, have more motion than the other atom, *b*, on their contact, a portion of its

motion would pass to atom b, and b, being now more
expanded than before, would have more free energy.
If now a third atom, c, was situated next beyond that
to which the motion passed, it would be attracted
by b, and would attract it more than before.   Atoms
b and c would therefore come to contact, atom c
would receive more motion, and would then be more
expanded and have more free energy than before.
In like manner the motion would pass to d, to f,
etc., each atom, in its turn, receiving more motion
by contact with the atom preceding it, and having
more free energy, and then, parting with its excess
of motion, and its energy, it would return to its normal state.   It will be remembered in the chapter on
force, (last pages) the explanation was made, as to
why in the expansion of any multiple atom, the
more released are the energies of its final units.   For
the sake of simplifying explanation in the foregoing,
we have supposed the exchanges of motion and of
primary force to take place along a right line; but
as a matter of fact, since each multiple atom is surrounded on all sides by other multiple atoms, the
exchange of motion and energy would take place in
every direction.   What we call magnetic effects concern more the entire mass of a body; while electric
effects are more manifested on their surfaces.   The
latter is generated by surface motion; the former
more by motion that affects the atoms of the mass.
But the explanation of the cause of each is in no
wise different.   In magnetism, the motion being more
diffused through the mass is more feeble; but Faraday obtained in magnetism, the magnetic (or elec-

tric) spark by simply intensifying the motion in the mass. Without intensity of motion superadded, magnetism is, relative to electricity, merely a static form of the same phenomena. They are mutually convertible, and both are produced by communicated motion. An electric effect may be brought about with far less motion than a magnetic effect. This is a natural consequence of its being more of a surface effect; the energies of atoms there, being more susceptible to communicated motion, and more responsive to it.

Electrical phenomena may be defined as manifestations of inequality of distribution of primary force or energy, imposed on the surface of matter by the inequality of atomic motion. In the separated surfaces of two masses, these manifestations are the differences of aggregate or resultant free energies disengaged by communicated motion. That of most energy would obviously be more intensified, and under a higher tension than the other, and would be plus, or positive, to the other; but it must be borne in mind that this expresses mere degree of force and nothing more. If *these surfaces* were placed in contact, the excess of motion of one would pass to the other; and with it, from multiple atom to multiple atom, successively, would come the equilibration of force, in the same manner as has been explained for the line of atoms *a, b, c, d,* etc. Of course to our senses, this motion would pass as one flash, or spark; —and doubtless to this impression given, is largely due the idea of an " imponderable fluid " passing. We cannot realize vibrations of multiple atoms of

matter, so rapid that twenty-three millions of them
occur in a single second, as in the production of some
colors; and yet the vibrations of the final units of
each atom must be far more rapid than this. The
degree of expansion required of multiple atoms in
order to produce the most ordinary electrical manifestations, is incomparably small:—and by a very
little frictional excitation, small light bodies take up
movements of attraction or retrocession. For example in the decomposition of a single grain of
water, or the separation from union of its oxygen
and hydrogen, the energy overcome is equivalent to
that manifested in eight hundred thousand electric
discharges, each of which is represented by the
electricity furnished in thirty turns, of a powerful
plate electric machine. If a part of the surface of a
homogeneous body,—or one whose multiple atoms are
alike, as in elemental matter,—be frictionally excited,
there is heat generated, (or the multiple atoms swing
in greater arcs, but since the atoms are alike, this
heat motion is rapidly diffused through the mass.

There is doubtless a small electrical effect, but it
is wholly masked by the far greater heat motion. In
heterogeneous bodies, the multiple atoms, or molecules, are unlike, and if heat motion be communicated by friction, it is with far greater difficulty
diffused or transferred, and the multiple atoms, compelled to take up, in some way, the motion received,
must do so partially, at least, by expansion. Here in
each atom expanded, and having more motion of its
units, there is a corresponding disengagement of
primary energy, and the general character of the

phenomena is electricity. As a familiar illustration, one with dry shoes may slide on a carpet, then touching a finger to the stove, there follows the electric spark; or, if to the open gas jet, the gas will ignite. It is therefore not only motion transferred to the gas, but motion of a character so intense, as to set the multiple atoms of the gas (so highly susceptible to motion) into heat vibration. Yet the motion, transferred to the gas jet, is not heat motion, for it has not the effect of heat to sensibility :—and yet the original frictional motion between the feet and the carpet did generate a great deal of heat. As stated in the subject of heat, it is an oscillatory vibration of multiple atoms, which expands the volume of the mass. But in neither electrical, nor in magnetic effects, is the volume of matter enlarged. A bar of iron may be magnetized by hammering,—this creates friction among its multiple atoms, and some heat ; but if there was no other kind of motion, except heat motion, there would be no other effect manifested.

What then is the probable nature of the motion producing electrical and magnetic effects? It is obvious, that though there is motion, it is something besides the oscillatory motion of atoms in heat ; and it is further obvious, that this motion is connected with a development of a form of energy, different, in its manifestation, from heat energy. In elemental or homogeneous matter, where the molecules, or multiple atoms, are all alike, the only other kind of motion of the atoms that could be generated would be a motion of their composite or final units, giving expansion to the molecules, or multiple atoms, and this is manifested as either electricity or magnetism. In heterogeneous matter, where the associated molecules are unlike, and where heat motion is communicated, either by rubbing, or by shock of impact, or by radiant heat, this motion cannot (for mechanical reasons) be diffused as fast as received; it must therefore, as stated above, result in the expansion of the molecules, or multiple atoms, and in the manifestation of electrical or magnetic effects.

Electricity as a dynamic effect, cannot be transferred or communicated, without a material medium of contact, any more than any other form of motion. " Experiment has shown, that a certain portion of matter, though it may be attenuated to an extent beyond the limits of calculation, is necessary for the transmission of the electrical discharge." " Induction, too, must take place through material atoms so near each other as to be within the reach of each other's attraction." " The theory of Faraday," remarks De la Rive, " rests on a sound principle, that electric actions take place through the intervention of material particles; and it tends to bring electric force into closer connection with other natural forces." And Grover says, " The gradual accumulation of discoveries is rapidly tending to a general dynamical theory, into which that of imponderable fluids promises ultimately to merge."

Enough has been said, to satisfy us that this whole subject depends simply upon the relations of motion and primary force. The common experiment of the pith ball being alternately attracted and repelled by the prime conductor of an electrical machine is a repetition, on a graphic scale, of what has been described as taking place between the multiple atoms *a, b, c, d,* and *e,* the atom *a* having the greatest motion, and its communication successively to the others. The pith ball is attracted, because its resultant of free energy is less than that of the conductor; there is contact,—tendency to equality of distribution of energy ; then repulsion to a position of equilibrium.

This is followed by the ball parting with some of its motion (and some of its resultant free energy). It is then in the same condition as at first, and is therefore again attracted and repelled, and this may be indefinitely continued. *All interchange of motion, and of primary force,* in matter, *is in the direction of most developed energy, toward matter of less developed energy ;* and as the most developed energy of primary force, is where there is the most motion, it follows that the direction of interchange of motion, and primary force, will always be the same.

MAGNETIC MANIFESTATIONS AND THEIR EXPLANATION UNDER ONE PRIMARY FORCE.

The current explanation of a magnet is, that " It consists of a collection of particles, *each of which is magnetic,* and endowed *with both kinds of magnetism.* In the unmagnetized condition of the mass, these forces are mutually combined, and exactly neutralize each other; but when the mass becomes magnetized, the two forces are separated from each other, though without quitting the particle with which they were

originally associated. All of the same kind are then
disposed in one direction, and all of the opposite
kind in the other direction." Surely there is nothing
equivocal or doubtful about this language. And the
logic of it plainly is, that in each final unit of matter
are located three distinct inherent energies, to wit:
two kinds of magnetic force, and the attractive force
belonging to all matter; for the assignment of these
two forces to the particles is necessarily their assignment to the final units of matter of which the particles are composed.

If more is meant by two kind of forces, than
forces different in intensity, direction of action, and
quantity, it should have been expressed, for this is
the limit of any ordinary conception of difference of
forces. The explanation above quoted goes on to
say: " Each particle thus acquires a polar condition, and adds its inductive force to that of all
others; as a necessary consequence of such an arrangement, the opposite powers become accumulated
at the opposite ends of the bar." In other words,
at one end of a magnetized bar are accumulated
attractive, and at the other end, repulsive forces.
And yet strange to say, these two opposite energies
attract each other! Notwithstanding that, any inherent energy must have sympathy with, and association for, its own energy, more than for that of an
opposite kind; otherwise it would not only be self-
dispersive, but the asserted opposite powers in each
particle and mass would never combine to act, as
they are asserted to act,—those of the same kind together. If this be true then, that the same energies

must have the same sympathetic action, if at one extremity of a magnetized bar be accumulated the attractive forces, they should draw toward them the attractive forces at the end of a second magnetized bar. Instead of this, however, they draw the opposite or repellent end. This theory then of magnetism, of the mutual association in the same particle of these two distinct energies, together with a third attractive energy, is not only inconceivable, but is, in itself, contradictory. But if we assume, as we have done, that all matter has but one primary energy, than the basis of magnetic, as well as of electric phenomena is inequality of motion, expressing, as it always does, a necessary difference between the primary force free to act ; either between two or more multiple atoms, or between parts of surfaces, or of masses. In magnetism, these differences of intensity are, as in electricity, styled polarity ; and for the sake of avoiding confusion this term will be continued. Electric polarity may be excited in all substances, as may magnetic polarity. All discovered properties of electrical and magnetic polarity indicate that they are merely degrees of the same form of development and induced by the same causes : viz., a vibratory motion, communicated as friction ; by hammering ; by alternately magnetizing and demagnetizing, as in the swift revolutions of a motor, which sends along a wire inconceivably rapid vibrations ; in each case the multiple atoms affected are expanded, liberating energy. Their action under such impulses has been already fully explained, and it is not desirable to repeat it. In electrical manifestations, the action is small, and naturally affects the surfaces most, since surfaces present the least resistance to motion ; even in ordinary magnetic manifestations, they are chiefly of the surface. Prolonged vibratory motion of great quantity, however, influences finally all the atoms of a mass, and the phenomena are then magnetic. Motion in either surfaces or masses, subjected to incident force, would from various causes meet with greater resistance in some parts of the surface or volume than in others. In heterogeneous bodies there would manifestly be greater resistance to a surface motion in some parts than others, because of the different constituent elements ; the motion would with difficulty penetrate the interior at all, and would be far more intense at the surface and would therefore appear as electrical differences. The motion of one portion being more than the other, there would be in that portion relatively more free primary energy.

This whole amount of energy would be a resultant energy, and it would tend to act toward the lesser resultant energy of the portion of less motion ; on the fundamental principle of the tendency to equality of distribution of free primary energy in the matter. These two resultants (to adopt present nomenclature,) may be designated as poles of energy—a pole of plus, or more, and a pole of minus, or less energy. Precisely the same effects, and from similar causes, would be brought about in a *mass* of matter subjected to incident force—whether homogeneous or not—and there would be a plus and a minus resultant, or pole of energy. The earth itself is an illustration of this. In its daily revolution, the incident forces of light and heat from the sun create a vast amount of motion in the part exposed. This motion is a motion of friction of parts, as well as of intense vibration of those parts. But owing to the great heterogeneity of elements, this amount of motion daily imposed cannot be readily transferred to other parts, and there results not only the oscillatory motion of multiple atoms of heat, but their motion of expansion, producing electricity. And, of course, with rotation there is an electrical current around the globe in planes perpendicular to the axis, thus making a magnet of the whole earth, with a resultant plus, and a minus pole of energy. Of course, these poles do not, and of necessity could not, coincide with the axis of gyration.

Should large motor effects, in either electricity, or magnetism, ever be produced directly from heat, it will probably be through heat applied directly to either heterogenous masses or vapors.[1] Enough has perhaps been said to convince, that the hypothesis presenting an explanation of electricity and magnetism under the assumption of one primary energy, definite and identical in character in all units, is at least rational. Electricity and magnetism are mutually convertible, and as each is a mode of motion of matter each will produce heat motion, and each

[1] Through such a volume there must extend minutely ramified metallic conductors leading to one main stem :—just as in the heterogeneous animal body, heat from food is partly converted into magnetic and electrical effects, which are collected in the nerve channels, and in the brain.

may be produced by heat; and any kind of motion producing the one will produce the other. The action of each along a line of multiple atoms, or molecules, is the same. That is, a motion of expansion conveyed to the first atom; this expansion freeing a certain amount of primary energy; the atom intensely attracting its next neighbor in consequence of that energy, and imparting its own motion; and it, in turn, repeating the process, and the same along the line of atoms; the last atom attracting and discharging its motion upon any contiguous substance. If instead of one line, we suppose a vast number of lines of atoms, as in a conducting wire, there would be a wave of motion and of energy, from a disturbing centre of motion along the wire; the wave passing successively through the multiple atoms, and at the terminus attracting other matter and discharging itself, it may be, with a great shock. The immense amount of energy displaced in the development of these two forms of force, electricity and magnetism, arises from the release, along their path, of great stores of primary force, which has all the effect of a wave of imponderable fluid, though in fact a wave of motion and of force.

CHAPTER V.

CHEMICAL ENERGY AND THE PHYSICAL PHE-NOMENA OF VITAL FORCE.

Chemical action a Differential Attraction—No Differences of In-herent Forces of Elements—Apparent Differences Accounted for—Current Electricity—Chemical Action Releases Energy—Vital Force not Possible except under Conditions of great Hetero-geneity of Matter—Its Characteristic Phenomena—Chemical Action Renders it Possible—Opposite Action of Life, Energy, and Chemical Energy—Theological Conceptions—What we Witness in Life Force—Channels of its Energy—Its Nature.

IT is the province of chemistry to deal with the forces and motions of the molecules, or multi-ple atoms, acting between different forms of elementary matter;—or matter having different physical properties. Our design here is very brief, and limited to explaining for chemical phenomena, the principle of one inbeing energy in the final units of all matter, and its probable mode of action in chemical forms of force.

From observation that molecules of different ele-ments exercise a variable attraction toward mole-cules of other elements, has been derived the term "affinity," signifying an elective tendency analogous to the likes and dislikes of sentient beings. This

conception has greatly fortified the imagination, by
investing each elementary form of matter with something more than mere difference of force, and it sug-
gests a hypothesis utterly unthinkable when applied
to the only properties of force known.   Even assuming, for the sake of argument, that the final units of
each elementary substance are endowed with some
form of property, or energy, unknown to all the
others,—and other than intensity and direction of
action,—such as is embraced in the words "elective
affinity," we must still deal with a fixed character in
each final unit of substance, whatever that character
may be.   And having done this, that it should then
be asserted that the same unit is invested with a
force, which can exert itself alternately to repel, and
then to attract, or *vice versa*, is inconceivable.

We are bound to exclude admissions of the joint
occupancy of the same unit of matter, by two or
more distinct energies, with first one, and then the
other dominant ; and we must conclude that the
variable action of molecules in chemical phenomena
is a differential attraction, arising from some force at
liberty to act, or prevented from action, by employment in other efforts and directions, the same as in
the case of any motor whatever.   Chemical action,
heat, and electro-magnetic currents are mutually convertible.   Attraction between multiple atoms (molecules) of different elements will always take place,
where the developed and unemployed primary force
of those of one is greater than those of the other.

The atoms of elements assimilate to the electrical
and magnetic, positive and negative states already

explained, and present the same orders of phenomena. As has been elsewhere said, in the ratio
that the final units of a multiple atom are separated,
is its primary force free to act, being relieved to that
degree from a concentrative exertion between the
units. Since the final units of any two elements
differ in magnitude, the normal atomic expansion of
the multiple atoms of each is quite different; and
the relative amount of primary energy of the atoms,
free to act outwardly, is different. Hence the multiple atoms of one element may present a negative
condition with respect to a positive condition of those
of the other. If these atoms be brought within the
range of each other's influence, (the two elements,
of course, being disintegrated, or in a state of solution,) the tendency to interchange of motion, and of
union between them, is, other things equal, in the
ratio of the difference of the free primary energies.
If the atoms (free to move) of three or more elements, be brought within each other's influence, the
tendency to unite is strongest, other things equal,
between those two whose atomic difference of free
primary force is greatest; though degrees of motion, created by heat, may differ so widely as to
determine a union in other directions. Since the
free primary energy of a multiple atom is greater in
the ratio of its expansion, atoms of large units will
have relatively less free energy than those of small
units.

Therefore, other things equal, substances at the
extremes of the scale of densities and specific
gravity, should unite with rapidity. But here the

9

mechanical difference of vibration of their multiple
atoms may prevent union, as in the case of hydrogen and the metals. Between other elements
the chemical bond of union may be very slight,
and a small agitation, as a limited heat motion,
easily separates them. It will be remembered that
in the interchange of motion, it always follows
paths of least resistance ; and for that reason alone,
as will be readily seen, the interchange between
multiple atoms would be very far from always being
the most active between those whose final units
were of greatest, and those of least magnitude. We
have spoken of chemical energy as one of differential attraction ;—an attraction of different degrees of
intensity between the atoms of different elements.
The cause of attraction, manifested between atoms
of certain elements, and of indifference between
those of others, has been partially anticipated. Of
the atoms of two elements, if their atomic, free primary energies were nearly equal, no union would take
place ; on the contrary, a mutual repulsion would be
possible, if the atoms of both elements are from any
cause much expanded, such condition of the atoms
corresponding to an intense "positive" state ; the
repulsion being due to correspondence in amount of
motion of the units of the atoms, and their equality
in free energy. Likewise a form of repulsion due to
excessive motion, might arise in the co-mingling of
unlike atoms, unequal in their development of primary energy. For the interchange in this case would
be very energetic,—almost instantaneous ; and the
whole effect of motion being generated almost si-

multaneously, the result would be a manifestation of great energy, even of explosion. Regarding multiple atoms of most free force as " positive," and those of least, as " negative," in all combinations, we have an explanation of chemical phenomena upon the simplest possible principle, similar in every respect to those of electrical and magnetic manifestations. The negative atom passively invites the interchange to the same degree that the positive atom solicits it, whenever the atoms are within each other's influence. Here are opposite states ; reciprocity of action, equal, and in opposite directions ; a union of the atoms ; and a mutual neutralization through interchange of motion. The union will be permanent in case the natural differences of magnitude of units, and of their motion and free energies permit.

The transfer of the unit vibrations of atoms, and their interchange of motion and of primary energy, from atom to atom, comprehend the phenomena of current electricity, from chemical action. If we suppose a battery prepared, the interchange sets out from the more oxidizable metal, and it traverses the liquid toward that less oxidizable, which acts as a conducting plate, and from which may extend a conducting wire of indefinite length. The amount of energy is in the ratio of chemical action, the direction of current, depending on the direction of chemical action. Assuming now a single line of the similar atoms, one of them at the extremity of the line receives the interchange of motion of an atom differing from it, and having relatively more motion and free energy ; this atom dashes against the ex-

treme atom, setting up motion of its final units by
interchange, and developing its primary force. It, in
its turn, interchanges in the same manner with its
next neighbor of the material line, to which its relation in the sense of motion and free primary force
is positive. Each atom impinges in like manner
upon the next succeeding atom throughout the line,
and a wave of motion passes, accompanied by what
is, in its final effect, equivalent to a wave of free
primary energy, which upon arriving at the last atom
of the line is ready to interchange further, or to
manifest itself vigorously as a powerful attracting
agent. Obviously, in this primary, or pioneer wave,
the interchange can be only partial. It is probably
instantly succeeded by others, and these again by
others, until the equilibration by interchange is complete. Each atom after interchange with the succeeding atom, repels it, and is itself repelled, since
both atoms are, with respect to each other, then,
" positive " in the relation of motion, and of equally
developed primary force. This action would give
rise to a small reverse current.

In the interchange of motion and attractive energy,
we find the explanation of the nascent energy of
matter, so prominent in chemical action ; for the
energy of an atom is proportionate to its free primary force, and that, of course, is greatest when it
is most liberated, which is by the expansion of the
atom ; and this expansion will be greatest when the
final units have the most motion, and this is immediately after motion is thrust upon them by chemical action. Chemistry reveals to us the vast amount

of primary force that is bound up in the union of atoms into a mass, the same as that required to rend them asunder. For example, the amount of heat motion required to raise one pound of water one degree in temperature, would, if mechanically applied, raise seven hundred and seventy pounds one foot high. The fusing point of iron is little less than three thousand degrees Fahrenheit. How vast then is the mechanical equivalent of heat motion necessary to melt one pound of iron!

These forces bound up in matter are, to our senses, latent, and are employed in holding substance rigidly together. The separation of substance into its multiple atoms, by imparting to them more than normal motion, is the release of a vast energy; and the expansion of those atoms, by conferring a greater motion upon their final units, releases another amount of energy equal to the force consumed in giving to their units a greater motion of expansion.

THE PHYSICAL PHENOMENA OF VITAL FORCE.

It is not our determination to enter upon the problem of organic phenomena, further than to arrive at their general relation to other modes of motion, and their manifestation of primary force. Here, as in other forms of physical energy, we concede to matter but one inbeing energy; and that the final units of substance differ, in each element, only in magnitude.

Chemical action is the first departure to heterogeneity from homogeneous elements; these elements, it will be remembered, being mechanically seggregated in the cooling of universally diffused matter. From chemical action to those modes of motion of matter, and manner of development of primary energy, which are favorable to the action of vital force, matter attains conditions of heterogeneity to an extreme degree. Highly complex molecules, compounded of atoms of many different elements derived, it may be, from double, triple, quadruple, and higher degrees of chemical union, in breaking up present us with our first ideas of the immensity of the energy in matter.

The physical phenomena of vital force are the incessant dissolutions and recombinations of highly complex unions ; primary force being here perpetually released, made continuous in flow, and economized ; as in a magnetic machine, a continual effect comes of magnetizing and demagnetizing. The way for organic life is prepared by chemical action in creating the great heterogeneity of matter and in the resulting rapidity of its interchange of energy. And conversely, no life will be initiated in matter, until these conditions are brought about. Without these immeasurably swift transformations life conditions are unfavorable, and in life manifestations, in the ratio that they are retarded, will organic activity stagnate.

The *free action* of primary force, is the *physical characteristic of life force.* Chemical action is assisted by heat and light motion, and by electric and magnetic motion ; together they heterogenize matter, pulverize and inter-diffuse it. They prepare the way for the latent life principle, which, like the life of the seed, is dormant, and remains so until conditions allow it to act. When it does act, the province of these motions is to tear down, as life force builds up. There are a few, even of those who accept universal law as the method of the universe, who are ready to concede that vital activity is a resultant, derived from the other physical affections of matter. The direct creation of organic forms, to most minds, and to others, their modification by surrounding conditions, is a satisfactory solution of the question as to the manner in which organic life has appeared; vital energy being presumed to be a quality of force, especially conferred for an occasion. Other thinkers, accustomed to dealing with experiment and laws, are divided between the germ theory, and that of the spontaneity of life under conditions of law. Commonplace theological questions, perpetually recurring, hold in many minds the prominent place, to which everything else is subordinate, as though the dignity and sublimity of the Infinite Being were at stake. Discarding the element, time, the controversy is resolvable into two questions; one of law, absolutely unchangeable, through the persistent operation of which, organic life is a possible development from the nature of force and matter as they now exist. In other words, that as they are fashioned, they are quantities and completeness, in themselves, sufficient to accomplish all the possibilities of Nature. The other question is one of modifications imposed from time to time upon processes of matter, by Divine interference.

Under either view there is nothing inconsistent

with the recognition of co-eternal, absolute Intelligence; and the theological difficulty arises, when
each class of advocates attempts to adapt the other's
scientific doctrine to his own theology. It will
hardly be contended, even by the most zealous
advocate of vital force as an independent energy,
that it is one non-resident in matter; or that its
being is possible unless so located. Not only is
this inconceivable, but if such existence were possible, it would most times be purposeless. If it be
a located property of matter, it is possible to all
matter; for all, or nearly all, known elements may
be assimilated in the organic economy, as useful
factors. The only remaining inquiry relates to *the
time* of the endowment of matter, with this supposed
distinct energy. Is it a quality conferred upon matter, in the ratio of the demands of the organism, as
indicated by its growth? If so, here is at once and
forever an end to all law and prevision, for there
could be no more complete assertion of interposition,
and no more sweeping denial of the truths of every-
day observation. Lastly, then, is vital force introduced for the first time, when geological conditions
are favorable to the development of organic life, as
organic germs? This idea is one involving increase
of primordial forces, and of special changes in them,
as radical as, though more infrequent than the last
supposition, and it upsets the operation of law,
since the energies of matter are, to our understanding, fixed, and whatever they are, that they remain;
at least until all matter under existing provisions of
force and motion is again diffused into its final units:

or until it stifles all possibility of life in sidereal con-
centrations, immeasurably vast and solidified. (See
subject of matter.) If the foregoing conclusions be
correct, the alternative is, that all forms of force are
derived from pre-existing form and quality; and all
forms of matter, whether organic or not, are potential somewhere in the nature of its own energies.
Under this view, the processes of the organization
of matter, and advancing progression of life forms
under Omnipotent supervision, are no more inconsistent with law, than are His co-existing intelligence, and sympathy with all His laws and processes
of law elsewhere in the universe. Conceding, then,
that the inherent force of matter is immutable, what
do we view in the development of organic structure,
and also in the vitality of that structure? Is it a
distinct energy until then impassive? If so, we
must concede to matter the association of another
kind of force to be added to the motley multiformity
of forces, which prevailing doctrine has conferred
upon it. There is then no alternative, but to admit
that potential life force *is a quality existing primor-
dially in the very nature* of primary force itself; and
as much a part of its character, as is its disposition to
attract other force from which it is separated, in the
separation of units and of masses. If this be the
case, then in organic forms, what we witness is a
manifestation of the life quality of the innate primary energy of matter, joined to the physical affections of light, heat, magnetism, electricity, and
chemical action, which are the same physical and
mechanical modes of energy, persisting in operation,

and in the same manner, as we have seen their operation in inorganic substance. So that all that is required is *structure,* capable of receiving, directing,
and converting these forces into aids, in the assimi-
lation of extraneous matter. The definite development of the structure is initiated and carried forward
by life force.

Of course, the most sanguine could hardly expect, in organic forms, to trace these various physical forces to their exact equivalents as vital results.
Yet even in this direction something has been done
to show that the progress of life form is greatly assisted by them.

As before stated, chemical action aids in tearing
down and removing life structure no longer wanted.
Heat and light motion promote rapid interchange
of force.

" Liebig, by measuring the amount of chemical
action in digestion and respiration, and comparing
it with the labor performed, has to some extent established equivalent relations." " M. Helmholtz
found that the chemical changes which take place
in muscles, are greater when these are made to un-
dergo contractions, than when in repose ; and that
the consumption of the matter of the muscles, or
in other terms, the excrementitious matter thrown
off, is greater in the former than in the latter case."
" M. Matteucci ascertained that the muscles of re-
cently killed frogs absorb oxygen, and exhale car-
bonic acid, and that when they perform mechanical
work, the absorption is increased ; and he calculated
the equivalents of the work so performed."

Electric motion may, as is well known, largely assist in restoring suspended animation. And even where vitality has ceased, its interchanges and waves of force, along nerve channels, may produce to some degree the similitude of life motion.

The cause of plant heat is to some extent chemical, and of animal heat, nineteen twentieths are the same. No living body can originate power; it can only convert the stored force derived chiefly from food. The force of the animal organism is derived from the vegetable, and that in turn from the mineral, through chemical and other modes of reduction and interchange.

In the exertion of every organ, force is consumed, and parts are wasted and dissolved ; and in every movement may be traced the operation of the laws applicable to any mechanical contrivance, definite degrees of change being connected with measurable consumption of force. Organic interchange is greatest in the direction of most motion, and like all other motion it follows paths of less resistance. And as the path of less resistance is that of less motion, organic interchange is in the direction from points of motion to those of less motion. Mr. Herbert Spencer says : " In the case of organic growth, the line of movement is in strictness the resultant (line) of tractive and resistant forces. . . . The shapes of plants are manifestly modified by gravitation, and every flower and leaf is somewhat altered in the course of development by the weight of parts. . . . From a dynamic point of view, natural selection is the evolution of life along lines of least resistance. The multiplication of any kind of plant or animal, in localities favorable to it, is a growth, where the antagonistic forces are less than elsewhere. And the preservation of varieties that succeed better than their allies, is the continuation of vital movement, in those directions, where obstacles to it are most eluded."

The genius of Spencer first pointed out that rhythm is a characteristic of all motion. This momentarily recurring fact, perpetually present to consciousness, had, until then, failed to receive the attention of consciousness. As the maintenance of one mental state would be unconsciousness, so the maintenance of organic energy in one direction, would induce stagnation of organic life in other directions. Incessant irritations, rhythmical or vibrating interchange, in all directions, make up the largest resultant in vitality.

Rhythm of motion is the inevitable consequence of vibratory atomic and unit motion. And in these directions is the exercise of most primary energy, and therefore the greatest opportunity for life force. Parallels are everywhere complete between those energies making conditions favorable for the appearance of life force, and those modes of motion styled the physical affections of matter.

There are other channels of organic interchange of primary force and motion electrical in nature. Through the respiratory system are supplied nineteen twentieths of animal heat. But we know that interruptions to the uniform transference of heat motion in a heterogeneous body produce electrical

motion, and that if these interchanges are long continued, they will establish for themselves paths of
motion, in directions of least resistance.

In the animal organism, therefore, heat motion is
constantly transformed into electrical motion, and
this motion has established itself in determinate
paths. The nervous system offers in its character,
structure, and by experiment, strong presumptive
proofs that it embraces the chief channels of the
conveyance of electrical modes of motion. It unites
the whole animal system by rapidity of respondence
into a unit of effort, and the ramifications of its
channels, are, in all respects, co-extensive with the
distribution of animal heat.

In the torpedo, or electrical fish, electricity is
most strongly developed where the nerves enter
the electrical organ. It is a characteristic of matter
in organic life, that it has increased freedom of motions, and consequent intensity of action, by matter
passing to states of heterogeneity from those of ho-
mogeneity, the instability of the association being
commensurate with the degree of heterogeneity.
The close association of multiple atoms in masses
locks up the primary force of matter; but in the
ratio that these atoms are free to move, do their
energy and action become individual, their primary
force free, and susceptible of becoming economized,
and employed in determinate directions.

Matter, through the operations of its physical
affections or modes of motion, passes upward from
the elemental to the compound stage, thence to the
vegetable, then to the animal, in advancing hetero-

geneity. The aliment of vegetable life is inorganic, that of animal life, organic. And activity in each is in the ratio of the heterogeneity of the supply. From the simple cell structure of the first and humblest type of life, to the complicated organism of man, involves a vast stretch of time and of manifold change, but who, looking down this long vista of development, can doubt the might of the wonderful force conferred upon matter, and that they together may accomplish the will of the Omnipotent through the gentle and patient operation of law. Surely the Infinite is not lowered by estimates of His methods through law, instead of personality. Immutability is an attribute of perfection, mutability, of imperfection.

CHAPTER VI.

EVIDENCE FROM THE WORLD OF MATTER AND FORCE OF UNVIERSAL SPIRIT.

Nature's Methods those of Simplicity—Consensus of Opinion—
Life Force and its Origin—Of the Laws of Attraction—Of the
Conservation of Energy—Variety of Elements—Nature of
Attractive Force—Likeness to Spirit Force—Presumptions of
Materialism—Cognition, Self Consciousness, and Reason—
Limits of Understanding—Habit, Instinct, Heredity—Sub-
stance and Spirit—Life Force and Life Conditions—Advance-
ment.

IN our investigation of matter and force, we have
endeavored to strip them of the occult and
mysterious. Upon their analysis we have
found substance, real, and of final units ; force dynamic, represented by motion ; and force inbeing,
represented in its aggregate form by the attractive
power of matter. If science in its steady unheed-
ing advance, with one effort, unpityingly drag away
the supports of the miraculous, and sweep aside doctrines of special agency, pointing in their stead to
undeviating law ; with another effort, it as ruthlessly
destroys the places of refuge of that bald materialism which says, " We do not know all the powers of
matter, its magical and spiritual nature, and its life

eternal," and which claims that an atom and motion explain all; that the universe of life is their product; and that " Laws are merely relations of things which exist of necessity." The examination of force and matter conveys at least the satisfaction, that by it we are in a better position to judge whether their physical forces are all sufficient, as physical forces, to produce life, self-consciousness, and reason ; or whether that inbeing force has a more subtle, inner nature, not involved in physical changes and phenomena, but without which life is impossible. We are not so inordinately conceited as to presume that the theory of one force, which we have presented, of the mode of production of the physical affections of matter is either the most correct, or the most probable. It is merely given as *a possible method* of their production.

But whatever the method be, it is sure to be very simple, and as the form of motion of any one of these phenomena may assume that of any other; and as the inbeing force of matter producing one, will produce the others, the oneness or unity of force may be fairly presumed. *Each* of these affections of matter is *a simple physical force*, mechanical in its effect and operation, and its mode of motion is always the same ; it always has the same method of working and no other, like a machine, hence the united action of these physical affections would produce a single machine-like effect and no other. The wonder grows that there should have been ascribed to matter and to the force represented by its motion, the power to produce life, consciousness, and reason.

Until a comparatively recent date, the forces of matter were enveloped in mystery, and supposed hidden qualities contributed largely to the presumptions of ignorance as to its possibilities. As to the oneness of force and matter, we quote from several authors to show the consensus of opinion regarding it. " The existence of a single elementary force of substance, from which by differentiation, transformation, and the adjustment of proportions, all the varieties and properties of matter are produced, is an hypothesis, to which the whole drift of contemporary science is bringing us nearer with every fresh accession of knowledge." " The unity of physical forces is the point on which science has its eyes fixed. Already it has been demonstrated that heat, light, electricity, magnetism, chemical attraction, are exhibitions of one and the same power acting through matter. That all these forces may be transformed into motion and by motion be reproduced, is now something more than an hypothesis. Hence the deduction that all physical phenomena have one and the same primordial agent as their original generator." The Duke of Argyll says : " Science, in the modern doctrine of the conservation of energy, and the convertibility of forces, is already getting a firm hold of the idea, that all kinds of force are but forms of manifestation of one central force issuing from some one fountain head of power."

" The reduction of all living forms to unity, that is, to the cell, is an indication that the vital agent is itself a form of the one primitive force, and thus physiology tends to unity by way of morphology.

And this reduction of organs to unity, is true for plants as well as for animals." The materialist denying all spiritual cause, is now compelled to account for the production of life, consciousness, and thought, by the *modes of motion*, or the physical manifestations of matter, for which, in neither observation nor experiment is there a scintilla of evidence. Neither of these forces, separately, has the nature of life force, and there is no reason to suppose that their united action can bring forth qualities and character not contained in the components.

Life force has never been evolved in the laboratory, notwithstanding the most studious and persistent efforts. Not even the faintest sign of structure has appeared ; much less the power that dominates structure presenting itself as life force. Denying God, the materialist declares that life proceeds from matter and motion. He proves neither that this is possible, nor that there is no God. We believe in the advantages of science to humanity, and from every source we welcome its truths. And we be-believe that these truths lead, as they have always steadily led, to the most rational and exalted, and the most fundamental conceptions of an Infinite all-wise-Being.

Regarding, then, the assumption as unwarranted, that the physical manifestations of force are sufficient in themselves either to originate or to maintain life, we turn to examine some of the evidence that there exists a supervising Intelligence, a Spiritual Power, through whom the inbeing force of matter derives its life-giving or spiritual quality. It goes unchallenged that the characteristic law of the inbeing force of matter is that it varies in intensity of effort directly as the mass, and inversely as the square of the distance. If we assume that there is neither Universal Spirit nor intelligent control in the universe; that nothing exists but force and matter; then it is self-evident, that whatever the mode of action of the inbeing force of matter is now, *that it has been*, in all past eternity of duration, and *that* it will be for an eternity to come. It is, therefore, not strange that from this standpoint matter and force should be undeviating in their operation; in fact have methods of undeviating law of action. But what is extraordinary and inexplicable is, *why* that law of attractive power should be the precise law of variation of the force that it is. For it is susceptible of mathematical demonstration, that the limits of possible variation of the intensity of attractive force of matter, consistent with orbital motion and the preservation of general harmony in the movements of heavenly bodies, are exceedingly narrow. Variations of this force according to any *direct* law of the distance, would bring about swift confusion and self-destruction; while if the attraction varied inversely as the cube, or any other *functions* of the distance, all orbital motion would be annihilated and impossible.

Therefore, had the law of the variation of the force of gravitation, as now existing in matter, been some one of millions of laws of variation, as it might have been, planet and world and sun formations would never have taken place and their matter would still have been a part of the inextricable confusion of chaos. We find, then, that our planetary system with its elemental and its heterogeneous matter, and the physical affections of that matter; its organic life forms, and all the beauty and the variety of our world, could not have been brought about, except for this particular law of variation of the attractive force of matter. If then, matter from all eternity *so far as we can understand* be invested with one of the very few laws of force out of millions, that would have accomplished the present world results; and if that matter constituted in this wonderful way so that its effects have culminated in appearances of the highest wisdom, enabling it to bring forth life, and thought power of men, why then, this is proof presumptive of the intelligent exertion of that matter. And it is, in fact, not only cause, but intelligent cause. A distinguished author says: "In tracing upward a chain of causes, if we stop at any cause, or force, or principle, that force or principle becomes for us God, since it is an efficient agent controlling the universe."

Considered by the mathematical doctrine of chances, there are millions of chances against so few, that they can be enumerated on the fingers of one hand, that a law of the force investing matter should be the precise law that would render planets possible. Is it possible, that matter is by chance so self-constituted? Is it not more probable that an Infinite Wisdom, above conditions and prescience, has selected the law which conditions matter?

Again, as has been said in the chapter upon mat-

ter, there is no principle of science more immovably
founded, than that of the conservation of the dynamic, or motion energy of matter. Though appar-
ently ceasing, it persists from form to form of motion,
absolutely indestructible. It has also there been
demonstrated, that matter acting under its law of
attraction, and this principle of the conservation of
energy, will sooner or later, though at some period
immensely remote, ultimately arrive at a stage of
stagnation, from which it has no power of self-resus-
citation ;—and that this is inevitable under any assumption whatever, as to the limits of matter in the
universe. As has been shown, this final period of
matter will be one of its consolidation, and of absolute darkness, or of its universal diffusion, and it
will involve the whole stellar universe.

Now whatever has been possible under the operation of the laws above indicated, in the tendency of
matter to any final state, must have been already
consummated during an eternal duration. Therefore, if the two laws of attraction, and of the conservation of energy, which now control matter, are
the same as those which have directed it in the remote past, all matter must, under their operation,
have arrived, in some past period, at absolute equilibration of motion, and stagnation, in the form of
either universal diffusion or consolidation. The
forces of matter which brought about these conditions would inevitably and forever perpetuate them.
And had there not been an intervention of some
Absolute Power above and outside of matter and
its inbeing attractive force, matter would now have

been in one of the two states indicated, of universal
rest ;—since matter could not, of itself, have changed
its laws of action or its inbeing force.   There is then,
upon any assumption, a period of progression, of
culmination, of decay and practical death of the universal material world ; and it can be re-animated to a
renewed progressive life, only through an influence
and Power beyond law and above it, and for which
law is insufficient.   Nothing in fact but Spirit, Universal and Absolute, can under the great laws of the
universe perpetuate the universe ; and in the helplessness of its old age restore it anew to a life-pre-
paring, and life-producing condition.

It may have been that from one to another of these
vast and supreme cycles of matter of the past, Divine Intelligence imposed upon matter other laws
and conditions of force than those now governing.
He may have willed variety in the supreme universal
method, like the variety upon our little earth and
planetary system in the elementary forms of matter,
and in the organic forms of life.   He may have willed
that force and matter should proceed so far, and then
change, to conform to other ends that He chose to
bring about.

An Unconditioned Power is not conditioned even
unto itself ; otherwise He is Intelligence, forever in
self-imposed bounds, and eternal non-interference
with His own laws ;—and wills not.   But *God wills*,
and unconditioned as He is, His method may be
methods of change, as well as methods of law in
each change.

It has been shown that conditions favorable to life

force, are those of greatly increased freedom of the motion of matter in its minute forms, and of increased frequency of interchange of motion and energy. These conditions give life tendencies their opportunity. But they would not be possible without heterogeneity of matter, or diversity in its elements. In other words, if there were but a single elementary substance, and if matter were all iron, or all sulphur, or all oxygen, there could be no life ;—as then, the conditions favorable for the operation of the life property of the inbeing energy of matter could never be brought about. How came it that this variety of elementary substance exists? If there were intelligent purposes to be carried out, this was necessary. If there were no purposes, then it was all accident, mere happening. Is it probable that this variety is a self-assumed prevision of matter? Or is it more probable that it is a forecast of wisdom from intelligent source? Again: Whoever declares that matter and its physical forces, are in themselves all sufficient for all effects that are evident to cognition, must account for the production of life. And if he deny that matter has but one inbeing force, and assert that the qualities of each elementary form of matter are due to peculiar forces singular to that element alone, it is pertinent to inquire, whence came all this wonderful variety of energies, of which, as we have shown, there is no possible conception nor explanation in the whole domain of physical science. There are about eighty distinct substances, and of course, if each substance be favored with a set of occult energies and properties, peculiar to itself, it

becomes those who advocate this idea, to disclose
the source of these extraordinary qualities. If matter has empowered itself, then matter must be wiser
than its inertness signifies,—and above necessity.
For necessity means simplicity, uniformity, sameness. Necessity is universality. What is necessary
to matter anywhere, is necessary to matter everywhere.

We now pass to the consideration of the character
of the attractive force of matter which draws together
its units as well as its masses. As is well known this
attraction is a power reaching through vast intervals
of space between masses, as well as through those
inconceivably small intervals between final units;
and in all cases it acts independently of a medium
of transfer. It is pure force, extending to, and
drawing to itself other pure force, located in other
matter separated from it by an absolute void. Yet
through this absolute void extends this wonderful
energy, a pure force existing without matter! Is
not this our idea of spirit power? A pure force
without matter? Only we concede to spirit power,
intelligence. Whatever theories one may entertain
about matter as a reality, we presume it is undeni-
able, to all, that there exists in all matter, within the
limits of our visual universe, an inbeing force of attraction; and that this force is exerted through the
voids of space and is independent in its action of any
medium. How came substance to be associated with
such a quality, or energy, wholly unsubstantial and immaterial? The two, the substance, and its indwell-
ing energy, are utterly without likeness or simili-

tude. One is the burden, the other the carrier. Is it from necessity, or from chance? Is it a self-assumption by matter or by force? Or is it an association originating in the wisdom of Omnipotence? Which is the most probable? If from this association come, ultimately, organic life, volition, self-consciousness, cognition, thought, reason, does it proceed from the inert, helpless dead burden, matter, or does it come from this wonderful innate power directing matter—a power which has the veri-similitude of spirit? Who has not witnessed the ordinary phenomena of transference of thought? Two persons thinking the same thought in almost the same form at the same instant of time; though each has no intimation of and no means of knowing the other's thought, which may be utterly foreign to any previous thoughts of either. Mind reading, mesmerism, and hypnotism, are familiar to all intelligent people. Here also, as in attractive action, are illustrations of immaterial energies of the will,—thought power,—passing through space,—absolute vacuo, from one mind to another, wholly independent of any medium of transference. Such transfers are not modes of material motion sent through continuous matter, but immaterial action. They are forces of thought conveying from one mind to another, simulations of the originating mind. This is spirit.

Now the subtle power of attraction which draws matter to matter is not of course a power of intelligence, but it has in other respects in its action all the likeness of spirit, and in this likeness it is linked to our idea of spirit. And is it not more likely that it is a process extending from Universal Spirit, than that it is self-ordained in its character, and in its association with matter? If, as has been seen, all the mechanical modes of motion, manifested as the physical affections of matter, persist in organic bodies as the same mechanical modes of motion; and if there, we find nothing changed in the nature of their action or character; what is the energy that causes every unit of every substance to tend toward organic structure and then to organic life? Whence come the volition, and self-consciousness, and thought that follow? Surely not from inert matter or from its mechanical properties of mere motion?

And if we assert that there is but one inbeing force of matter, instead of the vast multiplicity of one special force to each form of substance; why then, we must likewise concede, that life force is something more than merely attractive energy and motion of matter. For no one will be so hardy as to affirm that life is a product of motion and attractive force only. There must be then, somewhere inherent in the character of this force, a life-producing power, or quality, capable of action, when the conditions of matter are favorable for that action, but until then masked and inert,—just as in the plant germ there is the capacity for life, but that capacity is never exerted until solicited by conditions favorable. The first effect in the assembling of matter, is the preparation for life conditions through its various modes of motion or physical affections. As in assembling the parts of a machine, separately, they are unmeaning, but in place, they are proper parts of one form, and are seen to have functions. We do not hesitate to say, with Sir John Herschel, that "the force of gravitation is the direct or indirect result of a consciousness or a will existing somewhere."

From our point of view, it is a force from the Universal Spirit, delegated to matter and existing with it, under law imposed by the Spirit. Life force is the highest quality of the same energy. It also tends to attraction, to confluence, to unity, but more than that, it tends to organic form, to structure, and to life. In short, attractive energy, the investiture of matter, is an impulse of Divine Will. It partakes of the perfume of His Spirit, and its law is the expression of His wisdom. The lower and more material forms of this energy are mechanical in their effect. The higher forms are manifested in the organic, or life tendencies of all matter. Life, consciousness, and reason, proceed indirectly from Universal Spirit essence; and life forms are in their development, partly a product from this same source.

The fundamental conception of matter and force, outlined in this chapter, indicates, that unassisted and unsustained, they are as helpless to originate the universe of stellar splendor, as they are to develop a world of light and life; and we see pressing upon us from every side, the improbabilities and impossibilities of other presumptions.

Yet the materialist has robed matter and force with the garments of Omnipotence, and at the same time has denied to his counterfeit creation either purpose or intelligence. Admitting, as we must, that the self-existence of a Supreme Spirit is inexplicable, how much more inexplicable is the self-existence of matter and force, self poised, the latter with its wonderful self-assumed law of action, and its power to bring forth life and intelligence, though possessing neither of those attributes! The former with its self-acquired variety of elements. And the two together, wholly undirected, yet exercising and displaying all the wisdom of Absolute Mind! Fools talk about the all-resolving power of reason, and declare, that they believe not, what they cannot understand. And as by no successive reinforcement of thought from cause to effect, can they go back to the beginning of God, they assert His existence to be incredible of belief. Let us glance at a few of the limits of understanding. Can these unbelievers comprehend a past eternity of time, or a future time unlimitable ; or a beginning, or the eternity of matter? Can they understand life movement in the simplest form of the germ cell, or the growth of a blade of grass? Or how they raise a hand, or move a foot? Can they tell what is the I, that cognizes impressions from the outer world through sensation? Can they explain what it is that becomes conscious in self-consciousness? Can they explain how it is, that the inbeing force of matter can stretch across vast world intervals of space, and attract the force of other matter? Can they understand the transfer of human sympathetic intelligence? Can they even understand the initial motion of a body from a state of rest ; or the discontinuance of its motion? In both cases, all intermediate changes by increments must be passed through, and as they are infinite in number, infinite time would be required. The vibrations of ether substance convey to us the light and heat motion of bodies so remote, that thousands of years are necessary for us to receive it. Can they understand that if we pursue a right line in any direction we should never arrive at the boundary of a similar stellar universe?

How little is our knowledge! How limited our comprehension! We have only a narrow span of view, as from an island in the midst of space. Around us on every side is illimitable time, and extent, without boundary or shore: and beyond our outlook upon the stellar world and our own planet, we know nothing. Before one can justly assert that there is no God, he must have lived forever, have been everywhere, and have known everything. "One must first have become thoroughly conversant with the entire universe. One must have searched through all the systems of suns and stars, as well as through the history of all ages; he must have wandered through the whole realms of space and time, in order to be able to assert with sincerity that 'nowhere has a trace of God been found.' He must be acquainted with every force in the whole universe. He must be able to count up, with certainty, all the causes of existence. He must be in absolute possession of all the elements of truth, which form the whole body of the knowable anywhere. Else the one factor that he did not possess might be the very truth that would disclose God." In short, to be able to affirm authoritatively that no God exists, a man must himself be omniscient and omnipresent ; that is, he himself must be equal to God.

Suppose that upon some independent point in space, there stood the most capable of men, surrounded by matter in a condition of universal diffusion ; and that there were delegated to that mortal, supreme power to devise laws, to form worlds which should culminate in organic life and like man himself.   Let it be further supposed that he knew nothing of the existing laws of matter and force, but that his mission was to formulate them.   Does any one for an instant presume, that in a time equal to the eternal past, the problem would have been solved?   And yet the materialist concedes to matter far more than this!

Turn now to the phenomena of cognition, self-consciousness, and thought power, and assume that we are products of the forces of matter, with no presiding force to guide them.   We find that the elementary substances of the world, forming rocks, the air, water, and clods of the earth, etc., are declared to have become self-intelligent, since they are all embodied in the physical man.   They solve scientific problems ; they calculate eclipses ; they reason ; they become moral, even religious ; they make love ; they conceive hatred.   And the logical outcome of the argument of those who maintain that there is no original, higher quality in the universe than matter and its physical affections, or modes of motion, is, that each final unit of matter has independent intelligence, since matter cannot possess in aggregate form

what is denied to its individual components. " Whatever is *evolved* from matter must first have been
involved in it." Under this conclusion, the intelligence of man is the sum of the intelligence of the
final units of which he is composed, and all of his
mental powers are but the sum of the powers of
those units. In other words, " the human body is
maintained in its entirety and integrity by the intelligent persistence of its atoms. When their harmonious adjustment is destroyed the man dies, and the
atoms seek other arrangements and relations." And,
of course, what was the individuality is dissolved.
The atheistic argument from which this quotation
is made, does not pretend to account for the astonishing intelligence of the atoms. The presumption
is, that they are self-existing. One God is denied
and many are accepted.

Of course, this theory of intelligence of final units
is irrational in the extreme. And if parts of organic
forms of life and intelligence cannot convey to the
whole what they themselves do not possess, then we
must accept the deduction that there is a spirit relation in the inbeing force of matter, inherited from
Universal Spirit during the eternal residence of matter in the shadow of Divinity itself.

Let us trace the course of cognition as authoritatively outlined, and see if we divine any evidence
of a spirit element. "Suppose that we have some
sentient (organized substance) exposed to the impressions of the surrounding world. The sense impressions of these surroundings leave traces in the
sentient organism ; these traces, or structures of dis-

turbances of the organism, corresponding to the impressions, are preserved, and constitute a predisposition to revival upon a repetition of impressions of the same kind. The revival of the same feeling by the same traces, being again made in the sentient structure, recalls former impressions, because it is a mode of the same motion, again moving over the same path, and the organism more readily yields to that motion, or has less resistance than to a new order of motion. As often as this same mode of motion is repeated, the traces upon the organism are repeated, and the new impressions find a readier and more convenient path for their movement. In each case there is a revival of former sensation of feeling. There is, then, a recognition by the organism, of the particular mode of motion produced by the impressions, and at the same time there is revived a recognition of it, as being the same as former modes of like motions, producing like impressions. This is the beginning of memory in an organism." This recognition of a sense impression, as being the same as a former sense impression, necessarily involves a recognition *of* something, *by* something. Surely recognition is not set up and created by the motion, for that is a mere motion of particles; nor is it set up and created by any former trace of the same order of motion; for that is a recognition of the motion and of the trace. It is a sensation reproduced. And it might as well be said that the vibration of heat motion, and its resultant effect, are the same thing as memory or mind, as to say that the above described sense, or motion producing impression, originates memory *of itself.* A sense impression, and the final umpire which is cognizant of that sense impression, are two quite different things. What is this power that cognizes? That is of itself conscious. Is it mere matter of the brain? Highly organized gray matter, or matter to whatever degree organized, simply means matter highly responsive to motion. Is it then, that this matter, highly susceptible to motion, becomes conscious of a former motion, because it is now repeated in the same form? This would be placing matter above any known properties of matter. Is this power of cognition, or memory, a power arising from the aggregation of organic matter, and from the united action of all the physical forces or affections of matter? If so, we are conferring consciousness upon physical forces. And we are conceding to an assembly of the final units of matter, powers which we deny to the units themselves, for nothing can come from aggregates which do not exist in units.

But cognition, recognition, memory, self-consciousness, are certainly truths. And if their powers are within the limits of the brain, then there is a something there, higher than matter or motion, which notes their effects. This is power higher than any of the physical properties, or affections, since to none of them, no more than to the units of substance, do we concede self-consciousness, sentiency, or intelligence. The conclusion which we draw is, that the essence of cognition and consciousness is a spiritual quality of the inbeing force of all final units, which impels them to organization, and that with this or-

ganization there comes a resultant spiritual power,
a flowing together from all the unit components;
higher, stronger, and more potent, the higher matter
is organized. It proceeds from universal Spirit,
which has invested and clothed all matter with its
forces. It is a power above all others, a spirit
growth from spirit. And as it pervades the physical contour of man and all organized forms, its spirit
form is their form. For in its completeness it is a
unit of spirit power, self-cognizing, self-poised, and
a self-contained continuity.

Again. What are hereditary transmissions? Habits? Instincts? Let us admit the gemmule theory,
of the bodily transmission by processes of reproduction in animal life of minute organized structures
which, as do all germs, determine the final structural
form of the embryo, and the physical, and to a certain
extent, the mental character of the integral animal.

Let us also admit that habits, applied to animal
life, are invariably respondent states of the organism
to imposed external conditions. Habit, signifying
a disposition of the organism to repeat, periodically;
particular manifestation of activity. The external
conditions to which the organism is subject, are
those of force and motion ; and habits, therefore, are
modifications of organic motion, assisting to bring
about in the organism, particular resultant motions
or effects. But is this all in either hereditary traits,
or in habit?

Under the supposition, that all there is in mental
phenomena, is the product of matter and its physical affections, can any one, or all of these united, of

themselves, originate particular peculiarities of mind, in any man or woman at particular ages, known as hereditary traits? Admitting that structure repeats itself in progeny, it merely gives opportunity to the spirit qualities of the inbeing force of matter, which, acting through structure, follows the channels of structure.

So in habit, a particular structural mode is imposed on the organism, as effects of particular orders of motion long repeated. Habit is most prominent as we descend the scale in the orders of animal life. There, organization is lower, spirit energy is less developed and less controls matter and its material forces. Following to its broadest conclusions, our evidence of a spiritual element in the inbeing force of every final unit, which induces and assists organization and structure, and then presents itself as the life power of that structure, we must find in the lowest structure and life forms, the lowest spirit power, the most habit. Conversely, we find the greatest spirit element, and the least habit, the higher the organization. Here, the organic forces are diverted in particular directions, and are proportionally withdrawn from organic action in other directions. Governing habits are therefore inconsistent with the versatility, variety, and multiplication of powers of a highly intellectual man or woman, of an elevated type of life; and we should expect to find them predominant in the lower orders,—the more characteristically, the lower the type of the organism; until where the animal immerges into the vegetable kingdom, its long, monotonous, rhythmical intervals of habit assimilate closely to the greater stability of vegetable life.

If sufficiently prolonged, habit finds expression through inheritance as instinct, especially if its recurrence is dependent upon a degree of organic development, as, for example, of the organs of reproduction, which to the bird or animal dictate home building.   As the parent of instinct, we should expect to find instinctive life in the ascendency in the low orders of life, and an emancipation from it, according to the degree that higher orders were approached; or that mentality should bear an inverse ratio to instinctive capacity.   We should further expect to find hereditary traits to be those most closely allied to the propensities and passions; unless the subject were of a high, intellectual type; or of a highly organized mentality, and consequent spirit power.   In such cases the individual would be most likely to exhibit traits of an intellectual character, from similarly constituted progenitors.

As organized matter is of a high or low degree, so is life or spirit power more or less refined and more or less developed.   Organic life, both animate and inanimate, possesses, then, to different degrees the essential life or spirit force.

Tyndall says: " Abandoning all disguise, the confession I feel bound to make before you is, that I prolong the vision backward, across the boundary or the experimental evidence, and discern in that matter, which we in our ignorance, and notwithstanding our professed reverence for its Creator, have hitherto covered with opprobrium, the promise of all terrestrial life." Of course Tyndall means by matter, its activities, and not its inertness.

The whole question is, does matter contain any qualities of force except physical qualities? If substance be inertness, it cannot produce what it has not. If its physical forces can bring forth life and intelligence, then they can produce what they have not, and from nothing can come something. There is no mind nor will power in substance, or its physical forces. Nor have they that spirit power which can impress through a void,—other spirit. Mind and self-consciousness are not an outcome from such sources. Substance has ever resided in the presence of spirit, and mental and spirit force are growths from the shadow of spirit, an inherent quality of its inbeing force.

Chaseray says: "Let us distrust our imperfect senses. Let us not be precipitate in denying the quality of the human being, because the scalpel of the anatomist cannot reveal to our sight an eminently subtle principle. Man is not driven to annihilation even under the hypothesis of materiality." Cabanis, a great physiologist, admits that a principle or vivifying faculty is necessary to account for thought, and says that the contrary opinion cannot be substantiated. " Neither the primitive cell regarded as an elementary form of life, nor any principle known to science, suffices to explain life itself, or that power of action, which is in the living being at all the epochs of its existence, and consequently in the cell. In addition, therefore, to the material and sensible elements, there must be in it a principle inaccessible to observation, and it is this principle which is the agent of life; the impelling cause of vital motion, and of all differentiations. Nature is an organism through which the Divine life is ever streaming, and imparting itself to all organic forms. Nature is subject to change, to the limitations of space and of time, and to consequent imperfection." Universal Spirit exists in matter under the self-imposed conditions of law; and were we, ourselves, not amenable to law but only the subjects, perpetually, of special agency, we would be irresponsible puppets without moral attainments, or merit accountability.

Without Universal Spirit, nature appears a stupendous satire, grim and terrible in her sardonic irony. And in all her manifestations, merely a mother of freaks, bringing forth human kind as she would storms of the ocean, or the clouds of the air, to flutter awhile, and then pass away. Beings of mere physical enjoyment; without purpose; without aspiration. A universe of nonentities, with no useful result. Without Fatherhood and without sympathy. A happening, which may not happen again. A dissolving into everlasting nothingness, from which blind, terrible energies have brought them. If this be all, space would better have been empty extension; or matter have remained at eternal rest.

There is responsive action in the life force of structure, to favorable change of environment, as well as to the enlarged action of the physical forces of matter.

Purely physical action only makes conditions. And if spirit force be originally necessary to produce life, more spirit force is equally necessary to produce a higher life. Hence, when conditions of the environment are advanced from stage to stage, the spirit element of force the more responds. In other words, there is an increase of spirit element with each higher manifestation of life, whether of plant or of animal.

Structure, *of itself*, cannot tend to higher structure; or life, toward higher forms of life without outward preparation of physical forces, to which solicitation there is a response of spiritual element of force. Outward conditions are mere advantages of situation to the organism, and improved environment improves the opportunity for the life or spirit force of matter. Plant life is always responsive. Animal life,—man, to his shame be it said, is not so. If the tenant be low, degraded, animal like, he will degrade the palace as well as the hovel. He will not advance with improved conditions by which he may be surrounded. The sum of all this is, that as spirit element of energy must *be exerted*, in order to produce structure, in conditions favorable to life, and with the beginnings of organic life; so it must be more and more solicited, and more and more manifest, and more and more necessary in higher forms of that life.

CHAPTER VII.

EVIDENCE OF SPIRIT IN PSYCHICAL PHENOMENA.

Thought Transference—Somnambulism, Hypnotism, Etc.—Community of Thought and Waves of Thought—Inter-connection of Mind—Spiritual Side of Inbeing Force—Individual Responsibility—Animal Sympathy—Corollaries of the Foregoing—Summary as to Spirit Quality of Life Force.

THE most astonishing thing in our existence is that we exist. Surrounded by the inanimate we are animated. Enveloped by the inert, unconscious and nonsentient, we are self-conscious. To have formed ourselves was impossible. And the inert presents to us neither power nor probability. Hence we look beyond it, as man has always looked, for a higher and an intelligent energy.

There are certain subjective phenomena, that have done as much to impress civilized man with the idea that there is a difference between purely material force, and a spiritual nature in force, as grosser natural phenomena have impressed him in a crude and savage state, that there is an overruling something beyond and above him. These relate to the intercommunication of mind without using the signs of language ; to the spontaneity of thought, which is often as much of a surprise to the brain of the individual expressing it, as to those who listen, a surprise indicated by deep emotion, as laughter, tears, love, or anger; and lastly, to the phenomena of supersensual powers, exhibited in somnambulism, hypnotism, and mesmerism.

The outward acts of the somnambulist are the external manifestations of an interior corresponding power, independent of the physical body. A principle of intelligence which directs the movements, and sees and acts independently of the organs of outward vision. A spiritual force, while the body is unconscious. Bacon defines it as "proceeding from the internal powers of the soul." Transference of thought is a commonplace occurrence, and presentation of evidence of it here would be both tedious and supererogatory. A striking feature of the proceedings at the annual meeting of the British Association in 1891, was the boldness with which Dr. Oliver Lodge, F. R. S., in his presidential address to his section, ventured the suggestion that it was the duty of science to investigate mysticism; that its facts could no longer be ignored or denied. Careful experiments on thought transference and cognate matters had satisfied him, that a method of communication exists between mind and mind, irrespective of the ordinary channels of consciousness, and the known organs of sense. He expressed himself as convinced that thought may be excited in the brain of another person, without a material medium of communication. He says, " The relation of life to energy is not understood. Life is not energy, and the death of an animal affects the amount of energy no whit; yet a live animal exerts control over energy, while a dead one cannot." The mysterious way in which important news flies from one part of India to another has long been an unexplained matter of astonishment. It is an historical fact vouched for by unimpeachable evidence, that during the great Sepoy rebellion in the north of India, information of battles and their results would be known far to the south, hundreds of miles beyond the English lines, and would be current among the natives long before the same information could be sent to English officials at these same points, by lines of rapidly travelling couriers. The natives had no visible means whatever of obtaining this information, and the inference accepted at the time, and now, is, that the method was solely by the transference of thought. It is well known that their learning in mental mysteries is far more profound than ours. They have an advanced knowledge of telepathy or mind reading, such as we cannot understand or appreciate. And their knowledge and practice of what we call hypnotism, is greatly superior to our own. Many of their mental feats are wonders that cannot be explained, unless by the theory of some higher form of hypnotism, than that with which we are acquainted.

As before observed, the time has passed when established psychic phenomena can be scorned as unworthy of attention, or pooh-poohed away as insignificant and unmeaning. On the assumption that such phenomena are from physical forces, they are outside of all of our knowledge of matter, as to experiment or experience. Explanations involve the

acknowledgment of mind force, or thought transmission, or thought influence without language;
proceeding from an individual intelligence, and
operating independently of any known medium;
frequently through very considerable intervals of
space; and acting outside of, and beyond the brain.
Many great discoveries have been made independently and almost simultaneously. Among them are
the forecast of a planet exterior to Uranus by Le-
verier and Adams. The conservation of energy.
The theory of heat, and of gases. The doctrine of
natural selection. Spectrum analysis. The periodical law of chemical elements. The discovery of
ether. The invention and application of the cal-
culus There is no such entity as the "spirit of
an age," except so far as applied to the advancement and tendencies of a people. All the greatest achievements of mind are beyond the powers
of unaided individuals. It would seem that at
periods, a great wave of sympathetic thought, either
of a high or of a low order, swept over peoples, over
communities and individuals, exciting to frenzy and
madness, or to virtue and purity. Accord of thought
brings unity in effort and direction, and when gath-
ered and given impulse, it overwhelms individual
wills. There are periods of great religious disturb-
ances; of great political changes; of great scientific
advancement; of great military conflicts; of great
mechanical invention. This is sympathy of thought, a
multitude of impulses uniting in one vast irresistible
current. It is both concurrence and continuity of
thought.

"There are harmonious conditions of mind, familiar and well recognized as the result of identity of ideas. Each mind reflects the thought of the other. These are of daily observation. There is an immediate attraction of one mind for the thought of the other; an energy imparted by one, and a yielding to this force by the other. This may be extended by silent thought to many. If a large concourse of individuals, as in a great and close city, are highly moved by one energy, the sway of thought is mighty. Men are swept off their feet; are urged to action under an all-powerful impulse not recognized at the time."

It is an irresistible, sympathetic connection of the people, although they may be intellectually incapable of attaining the idea by their private understanding, or even perhaps of consciously apprehending it." Of course mental constitution and condition must be in a responsive state for this effect. Under such circumstances and in such communities, there is restlessness,—an inexplicable condition of nervous tension, and apprehension of a something to come;—then there is realization and possible understanding. What is the explanation unless mind force is a real force, a power capable of producing its effects through space, and independent of any material medium? Mind is inter-connected with mind whether consciously or not; with some minds far more sensitively than with others. One may, consequently, be affected independently of his own ideas, by virtue of this condition, even before the subject comprehends it. If this be so, and there be continuity of mind and inter-connection everywhere, we receive from every direction impulses of thought power, and mind impinges upon mind everywhere. We are inter-connected through heavenly spaces with spiritual thought, through earthly spaces with earthly thought, and there is even " Continuity between man's mind and the Most High."

It has been demonstrated in the most positive manner, that through short distances, one mind may impress another with its own thought, when both are in their normal condition, and bodily separated. Each mentality has its own peculiar mind energy, or mode of acting, and these are as numerous and as diversified as is the human face, or personality. It is a common occurrence that some sensitive persons recognize the presence, or the vicinage of other individuals, or of animals, in the most decided manner:—and evince such recognition, either by emotions of attraction, or of dread and repugnance, though such presence is wholly unknown to them through the senses. And it is an occurrence equally common, to have such recognition by the inner mind sense subsequently verified.

Thus thought reaches out to thought through the antennæ of mind force, extending as rays of light extend from a brilliant point in all directions. It is a spiritual and immaterial impulse, and yet by its action, it is a recognized energy. By it, the universe of thought is co-related and made one, through Universal Spirit. The more it is in harmony, and in accord in any community, the greater its effects. And the more spirit thought is in harmony with the

Divine Spirit, the farther such thought may extend, for then, thought impulse is not antagonistic to universal spirit, and is carried wavelike, in all its freshness, perchance to some loved one gone before.

Each individual is then a universe from whose centre thought proceeds ; and whose circumference of influence is inestimable, as the effects of his thought are never ending, but go onward and outward forever.

As it is sympathetic, or the reverse, we invite or repel the thoughts of others. If we indulge in vicious reflections, we supply such influence to others. If we revel in images of sensuality, we invite kindred company, and group around us vile and degrading emanations from other minds. Contrariwise, refined and exalted contemplation wins sympathy and strength from the pure, pervaded by like thought.

If there be a spiritual, as well as a material side to the inbeing force of matter, the spiritual, being jointly a possibility, spirit-influence, as well as more obvious energies, must be allowed a place as a factor in all results which unite to make up organic life. Tendencies of whatever nature, disposition, excitability, ungovernable passions, emotions, even morals, and all psychological phenomena enter silently into all mental processes as results of inheritance; as do, also, instinct, intuition, indefinite apprehensions, and unconscious movements. It may be said, that functionally, diversity of arrangement of the final units of matter, in different brains, brings about different orders of resultants. This admitted, it is equally true that spirit, or life energies, have had a large influence in effecting any given order of arrangement of these units, and therefore in producing any resultant. Hence, while for some of the iniquitous tendencies of every mind the individual is largely responsible, for others he is not so much responsible; since, by no fault of his, a certain degree of bias to propensities or passion are his neither by choice nor by avoidance. And it may be, further, that bereft of every advantage by enforced poverty and misery, he has never been able to acquire the will power of self-control. How important, then, is education, and how great our responsibility toward the poor and the wretched !

We are now in a position to view, somewhat more intelligently, the lower animal sympathy of species and of kind. Let it first be observed, that every animated organized being possesses, in his individual aggregate, an order of motion and of action peculiar to his organism. The whole vitality is its expression, and comprehends all the mental and physical processes. The persistence of this vitality perpetuates the identity of the organism. Sympathy is based upon identity, correspondence, or parallelism. It springs from those things which preserve identity. An animal as a whole, and one of a community of similar organisms, is simply a vitalized unit, having particular orders of motion, the same as an organized molecule, an organic germ of a particular kind and magnitude. There is therefore a synchronism of motion and action in the vitalized units of a community of the same kind. This synchronism signifies unison, accord, harmony of association, consentaneousness in action and in tendencies. And of course it embraces all that there is in physical correspondence, as well as likeness in the spirit element.

To a certain degree, any individual of the lower orders of animal life, as well as man, may impress the aggregate, making up its identity, upon fellow organisms. For in the aggregate primary force, and life force of an animal system, the imparting of energy, or modifying the energy of another similar aggregate, in other words the interchanging of energy, may be compared to the similar interchange of energy and primary force between two molecules. The energy of each is impressed with the likeness of the energy of the other. Of course with the difference, that the interchange between organisms involves the spirit or life quality. Animals of a particular kind group together, or " Birds of a feather flock together," not because each recognizes in the other the counterpart of itself, in outward form and appearance, but because the spirit instinct of each is impressed by the sympathy of a similar spirit force or instinct in the other, and this element, together with that of similar tendencies and desires, maintain the association.

Let us now examine some of the general corollaries involved in the foregoing statements. Life evolution has gone forward with earth periods, and has, from time to time, advanced or retrograded, depending upon the solicitation of life forces by external conditions. The general direction, however, has always been that toward higher and more perfected life forms, until at last we have arrived at man. Yet we have no reason to believe that the present man is the limit of perfection. The embryo of the man-child has inherited from organic forms less perfect in spiritual force; and that in turn from still less perfect life, and so on down, until animal form emerges into plant form. It follows that if man has spirit life, all organisms have spirit life. As that spirit is a spark from the universal spirit, all spirit life persists, and no life perishes. Matter, by slow growth in the fertile womb of Nature, gropes surely and steadily toward organic form;—a grouping of life energies. Then to higher, stronger groupings. Then at last to man. In each form is the spirit element. Spiritual power is then an inheritance. It may be, by cultivation, developed into great possibilities, or it may be dwarfed and smothered by abnormal growths. "The mind is narrowed in a narrow sphere and the spirit grows to its allotted spaces." Spirit, then, has origin, growth or extension, and maturity, so far as the organism will admit. It takes its organization, limitation, and character, largely from the physical body, and the spiritual growth and development may be said to be a refined product of the organism, inasmuch as that is the centre of life and spirit. "Man may, therefore, be said to have two organisms, that which falls under the cognition of the senses, and that which is the invisible life spirit."

Brain life is the organic seat of growth of spirit activity. In an inferior degree there are similar qualities in nerve life and its affluents, for life or spirit quality depends on degree of organization. From

this flows the explanation of nerve habit; nerve
instruction, as in the fingers of the pianist; reflex
action; insistence of nerve memory, as when nerve
life, deprived of any usual functional exercise, or
indulgence, sets up an irritating reminder, amount-
ing to memory. Now, nerve diffusion is so multi-
form and minute, that no point of the exterior or
interior of the physical body is without nervous
ramifications, and a needle point, placed anywhere,
will pierce them. If, then, the physical body be
stripped of tissue, blood, and bony skeleton, we
would have left a sort of phantom duplicate of man.
But this is the repository of the life, or spirit force.
It is the spirit manikin, as well as the channels of
life force. And if we could see mind contour, it
would necessarily take this life form and appearance.

The last summary of our proposition is, that the
life, or spirit quality of the inbeing force of matter,
is a diffusion of the universal spirit, which like the
light of the sun is thrown everywhere. Matter is
guided in its tendencies by this spirit tendency, to
organic structure, and to the manifestation of life,
sensation, self-consciousness, and intelligence. Spirit
is everywhere, and the material atom must forever
move in a spiritual atmosphere; like particles of iron
in a magnetic field, it is perpetually bathed in a diffusion of force of a particular kind. As matter as-
sumes higher and higher organic life forms, spirit force
is more manifest. Spirit tendencies and powers acquire more strength, and propensities and passion,
not only in orders of life, but in individuals of any

order, are more subordinated to that spirit strength. Spirit force enters into, and is a quality of life force, of all organic forms. Spirit of a low organism is less than that of a higher; and in the ascending scale to man, spirit ascends. It is more and more developed and dominant, and has more independent power. Further: The grosser and more sensual the organism, the more is it animal, the less intellectual; the more material, the less spiritual.

CHAPTER VIII.

RELIGION OF SCIENCE A RELIGION OF GOD.

Idea of Supreme Spirit a Duplex Conception—Religious Thought
and its Evolution—True Religion and True Science—Results
Unite instead of Diverging—Religion Common to all Men—Of
the Great Religions of Earth—Evolution of Great Religious
Leaders—Spiritual Phenomena Psychic—Duty of Science to
Investigate and Develop Truth.

WHAT idea of the Supreme Spirit do we acquire from the standpoint of force and
matter—unbiased by previous education,
and by the suggestions of the religions of the day?

From the laws of its grand aggregations; from the
division of matter into elements; from the great
cycles of change through which it passes; we have,
as has been seen, evidence of a Supreme controlling
Spirit.

From the phenomena of organic life, sensibility,
consciousness, and intelligence, we pass to the idea
that our own being has to some degree the spiritual
essence of the Divine nature, imparted through the
processes of organic life. This is the whole logic of
natural religion.

There are here involved two conceptions of a Supreme Being. The first is of absoluteness and im-

personality. The second of a sympathetic, near, and loving spirit. These are the conceptions that every thoughtful man and woman, whether conscious of it or not, has of the Divine existence, whether they have sat under the religious instruction of the day, or have climbed up by the steep and rugged heights of science to a knowledge of God. The one is of a Being Absolute in power and qualities, All Supreme, pervading all space and existing from all time to all time.

So far as this utter absence of relativity of ideas can be called a conception, it is an association in the mind, of a supreme omnipresent spirit of Intelligence, with limitless time, space, matter, force, and their undeviating law of action. We apprehend an unknowable extent of Spirit Being, so vastly beyond and above us, that nothing which we can comprehend of Him is much more than unsympathetic mystery and awe.

The second conception is more real, because more comprehensible; it is concrete in nature, an understanding of qualities applied to subject. We understand by it, Divine nearness to even Fatherly sympathetic spirit, and this is not absorbed in the first pantheistic conception. We understand God to be the director and supporter of all things in nature and in the visible universe; the origin of order, regularity, and beauty, and at the same time the Being who has given us vitality and intelligence. In this view the Supreme Spirit is associated with the humblest form of life. The tiny struggling flower of the desert, lifting up its pure, sweet face from the

arid waste, in thankfulness, is as much remembered
as the loftiest human personality. We remember
that the life and spirit power of each had the same
origin ; and that He has directed, and must share in
all the minute as well as the great operations of nature. We feel that He is in close sympathy with all
things, and that the wise and the weak, the lofty and
the degraded, are alike considered in His plans. And
though we recognize everywhere, law, both of thought
and action, we are impressed by a conscious sense of
the nearness of His spirit, and we know that our
purest and best ideas have a response in His spirituality.

In our relations to our fellow-man, we have the
sense of justice, of consideration, of sympathy, of
affection, of pity, and of love, and as we remember
that our life and spirit are derived from the shadow
of His being, we attribute to Him similar, but more
perfect attributes. We are convinced that He has
impressed some of the qualities of his own spirit
upon the life He has given to the inbeing force of
matter, for a purpose ; and that we are one of the
processes of that purpose.

It is this second conception of God that brings
about our idea of His nearness, and of our direct
relations to Him, and it does not at all involve the
idea of Divine interposition or of special agency. But
the best spirit of modern science does involve, and
as certainly necessitates and teaches, the obligation
to purity of life ; upright conduct ; obedience to the
highest demands of Divine law ; love of man and of
God ; and the closest sympathy with His spirit, as

seen through those wonderful phenomena whose assembly we call nature ;—as does what is understood
as revealed religion. The two point to the same end
and purpose, as well as to identity of effort :—that
is, elevation of man and love of God. *Hence the religion of science is a religion of God.*

Setting aside, for the present, considerations of revealed religion, we find that estimates of a Divine
nature, at any time, and among any people, have
been to a considerable extent reflections of contemporaneous enlightenment, or ignorance, and that
these estimates have been accordingly exalted or
degraded. We find religious thought and literature
to have been a development, not always progressive,
sometimes advancing, sometimes retrograding, but
a thought always modified by local influence of cli-
mate, not unfrequently with its songs of praise, and
basic ceremonies, influenced largely by the natural
resources of the country and of the soil, either spon-
taneous or developed. Nor can it, by any means, be
said that the religion of a people at any time was the
product of the best thought of which they were capable, or even that it was best adapted to their condition, but that it has always advanced, or failed to
advance, as a factor dependent upon the general
condition of the people.

Yet notwithstanding this, by the light of scientific
and material truths ; of general contemporaneous
knowledge, and by the aid of a gifted spirituality of
being, men have arrived, at different periods of earth
history, not only at lofty conceptions of a Supreme
Spiritual Being, but of the purest morality ; and

have attained great personal elevation of character, and the strictest practice of virtue. The Deity has, by them, been conceived to embody, in a perfect form, all those qualities with which we now invest His character, as goodness, power, wisdom, grace, justice, beauty, sovereignty, and love. The philosophy of their times has been pure and elevated, and their teachings closely approach to those of modern christianity.

We repeat that true religion and true science have a common thought and a common purpose, and that is, the highest good of humanity. They should labor harmoniously and in concert,—for true science leads to a broad understanding of God, and brings man into closer relations with His Spirit through nature. *Science arrived at the idea and conviction of a Supreme Spirit. The religion of science is a religion of God*, and differs not from that of revelation in its purity or practice. The aspiration of all good thought is mutual progress and advancement toward higher and better life conditions. As the elevation of man is the avowed purpose of religious instruction and of all religions, so the avowed object of all science and of all systems of philosophy, is greater happiness in this life, joined not unfrequently to the hope of a higher ultimate life.

All are aids in one work, and as such, are exalted expressions of man's conceptions and trust. The only absolute facts that science can present, are facts and laws of phenomena. When it draws inferences as to first or final causes, its status is precisely the same as that of any form of religion when doing the same thing. There are in both cases the same possibilities and probabilities. Drummond, the author of *Natural Law in the Spiritual World*, says: "All religious truths are doubtable. There is no absolute proof for any one of them. Even the fundamental truth, the existence of God, no man can prove by reason. The ordinary religious proof of the existence of God involves either an assumption ; argument in a circle ; or a contradiction. Entire satisfaction to the intellect about any of the great problems is simply unattainable; and if you try to get at the bottom of argument, there is simply no bottom there. All religion comes out of the literature and the thought of the time—progress. In the distant past there flowed among the nations of heathendom a small stream of religion, and now and then at intervals, men carried along by the stream uttered themselves in words. The historical books come from facts, the devotional books from experiences; the letters came out of circumstances ; and the gospels came out of all three. That is where the Bible came from. It came out of religion, and not religion from the Bible."

It is idle to prattle of a conflict between religion and science. If science has borrowed its moral supremacy from great universal truths, from ages of growth in ideas of utility and benevolence in the struggles of man, so religion has borrowed much and changed much, and acquired much from the truths of science. "Religion is not the acceptance or rejection of dogma ; it is a temper, a behavior." If science has in its discovery of truth, pointed to One

Supreme Spirit, no religion can do more. Churches and creeds, and religion, are not the same thing. The elemental idea of all religion is best expressed in the language of that lofty and lovely spirit, who spoke as never man spoke: "Thou shalt love the Lord thy God with all thy heart, and with all thy soul, and with all thy mind. Thou shalt love thy neighbor as thyself."

Creeds and dogmas are built up. They are the work of man, pure and simple. They come from organization, and the desire for strength, power, and importance. All the great religions, doubtless, have the same intention, love of man and of God, and His adoration and worship. But they prescribe many things more. Each form of belief, according to men, must have compactness, unity, and concert of action. From this idea have followed given forms, tenets, ceremonies, fast and feast days, all more or less pretentious and exacting. These are the built-up additions to the great fundamental principle above enunciated, alike the aim of science, and of all forms of religion. The heart and the thoughts of man; purity, love, unselfishness; the simple teachings of the loving Sermon on the Mount by the gentle Christ; and man bowing in the silence of his chamber to the supremacy of God; these are all in all.

All intelligent men, at all periods of time, have had a religion, and all religions conforming to the degree of advancement of the time have been useful. Each individual, whether professed materialist or atheist, or not, has a religion;—that is, a governing order of thought, or of principles,—rules of

conduct, that are to him more controlling, more intimate, closer to his personality, than any other
order of thought. Does not this universal groping
of mankind, this unconscious reaching out of aspiration toward the Supreme controller of things, indicate that deep within the nature of all, unknown
and otherwise unfelt by us, there is an unfathomable
yet profound natural sympathy with that Universal
Life and Spirit which we call God? "Through our
finite and fallible faculties we may not hope to comprehend God; yet science may lead us to even higher
and more rational conceptions of His possible nature.
It may help us to find truth both in the theistic and
pantheistic idea, and thus to reconcile what may
at first seem too antagonistic to be entertained
together."

The great religions of the earth, through a steadily
advancing enlightenment, have been constantly drawing nearer to each other and nearer to scientific fact
and discovery. The whole tendency of the movement is toward broader views of humanity and of a
Supreme Being. More charitable, more gentle, more
liberal to man; more fundamental, more universal,
more rational toward the great Fatherhood of God.
Still nearer they must approach, until all unite not
only in a common belief of a Father of love, but in
that of the brotherhood and kinship of man as derived in common from Universal Spirit; in one hope
of universal advancement, one belief in a spiritual
life hereafter and in one acceptance of law as applied
to our material world and life, as well as to the hope
of continued existence.

Science has neared the limit of its research as to origin and evolution of life. There is a limit in minuteness which it cannot pass ; and there is a limit in the beyond, which it cannot pass. Before and behind, inclosing all discovery and knowledge, are the unknowable and the unfathomable.

From time to time there have been born into the ages, though not of them, great personalities, whose teachings or works embrace not only all that has been evolved up to their time, but whose ideas and projects lead on to periods beyond them, spanning a far distant future ; ever seeming to connect with a period, when another incarnation of greatness appears, to take up the thought and work of the first, and to carry it on and on to ages beyond. Such men and women are rare, but of great powers, which we call genius. They have represented every path of progress : Religion, war, mechanics, music, sculpture, science, painting, invention, and architecture. They are the product of the spiritual and organic influences of the time. That such prodigies of humanity may appear, two influences must harmoniously conspire, viz.; spiritual sympathy of the age, and correspondent life harmony of the organism, with its environment. The spirit sympathy is from the united thought of the age, its wants, desires, hopes, aspirations. The life harmony is in unison with this, and they conspire to produce grand births. Great waves of religious influence and thought, dominating vast areas of the public mind, have produced marvels of mental and spiritual life ;—of men and women whose spiritual emotions ascended to

close sympathy with the Universal Spirit. "Per-
sonalties so far in advance of the great majority,
that their grandest mental conceptions are unex-
pressed, because thought symbols of language cannot convey the spiritual conditions of these few,
elected to walk as an advanced guard to generations.
They bear the outward semblance of the human
type, but until the last scar of mental and physical
disease, transmitted through the law of heredity, is
obliterated from man, must they walk in our midst,
—many times in obscurity, because born (spiritually)
before their time."

Such beings are in every sense inspired. Beyond
the sources indicated, " whence their emanation and
how they got their power we cannot see. There is
no explication of their lives. They rise from shadow
and go in mist. We see them, feel them, but we
know them not. They arrived, they did their of-
fice, with God's holy mantle about them and passed
away, leaving behind them a memory half mortal,
half myth. From first to last they are creations
baffling the wit of man to fathom." Mozoomdar,
the great leader of the Brahmo-Somaj in India, says:
" All the impulses and strong wishes for moral or
spiritual life become proved realities in these superior beings. The mind of God, faintly shadowed in
our hearts, kindles into a sort of supernatural light
in them. It illumines our path far before and behind. They become divine men. Not a few fall
down and worship them as Gods. The wisdom of
Greece took shape in a Socrates; the stoicism of
Rome, in a Seneca; the asceticism and self-conquest

of India in a Sakya-Muni ; its insight in a Krishna :
the Chinese sense of duty, in a Confucius ; the Arab
energy of faith in a Mohammed.   Each is a principle
of humanity, each a phase of the Divine reason, each
is a spiritual principle personified.   " God has not
left Himself without witnesses," without representa-
tives.   Such are the very best of earth,—men who
are sent in the double capacity of representing both
God and man,—they being human in the highest
and divinest sense, and divine in the most intelligible
sense."

Each of these lofty personalities belongs to his
own time.   Each represents its best thought, its
utmost aspiration, its necessity for change, and each
is the embodiment of its virtue.   The greatest of
all was Jesus Christ, above criticism, above imagination.   Pre-eminent love for man and God, utter
unselfishness, perfection in thought and deed, during a whole life, were all His.   He was fearless, just,
and patient, and with a gentleness of spirit that
belongs to God.   His was the living illustration of
Divine attributes, so far as they can be personified
in man's nature, and but for Christ they would have
remained abstract idealities.   The reflection of His
whole being is the exalted and pure Sermon on the
Mount, embracing all there is of morality, and all
there is in religion.   More beautiful words were
never spoken, higher, purer thoughts were never
expressed.   For the first time in the history of the
world, He taught that the religion of God is a religion of love.

There remains one other kindred subject which

has manifested itself as a religion of phenomena. It has attracted world-wide attention, and within the last forty years has attained proportions of thought in all civilized countries, which cannot be ignored. Comprehended under the general term of Spiritualism, it has aroused a degree of enthusiasm and intensity of interest among its advocates, equalled only by the ardent hostility and bitterness of feeling arrayed against it. On one side are millions of men and women earnestly contending for the purity of its doctrines, and the genuineness of its marvels. On the other, opposing such sentiments, are the learned ; the world of fashion and pleasure ; the intolerance of established forms of religion, and the men of science. Its believers are without organization, without leaders, and without concert of action ; the phenomena witnessed are not classified, and honesty is not always insured. Victimized on every side by charlatans ; imposed upon by the lies and fraud of pretenders ; fed upon by parasites, whose sole purpose is gain, it has been unjustly and unfortunately exposed to ridicule and contempt. Those who have the most readily accepted this form of faith are the humblest and uninformed, without the advantages of learning to enable them to judge wisely, and who yield to their great desire to witness phenomena, either to fortify a waning faith, or to gratify intense curiosity. Examined by the cold light of reason, we find that its phenomena point to psychic force, accompanied or directed by uncommon and surprising thought manifestations. And its philosophy we find to be what the broadest religion teaches, love of God and of humanity; that there is no royal road to heaven; that unselfishness, purity of conduct and thought are the standards for all, and by which all advance, here and hereafter. In many ways spiritualism has accomplished vast good. The investigation of all the phenomena of this subject, science owes to itself and to mankind. If science be a leader of physical thought, it should be a leader in all forms of physical and psychical questions.

Here are phenomena far out of the ordinary character, and vouched for by unimpeachable testimony. Surely if nothing be too lofty for scientific investigation, nothing should be too humble.

To assist those who are unable to inquire for themselves should be the purpose. If there be truths, let them be classified and added to the cabinet of scientific knowledge. If there be illusions, or baseless trumpery, or phenomena connected with laws already known, let their character be established. Truth alone can make man free. Perchance higher spiritual truths stand without, waiting to be recognized. All additions to psychic facts are links, binding together organic life and spirituality, and binding all individual spirituality to the Infinite and Universal Spirit,

> " That God, who ever lives and loves,
>   One God, one law, one element,
>   And one far off divine event,
>   To which the whole creation moves."

THE END.

www.ingramcontent.com/pod-product-compliance
Lightning Source LLC
Chambersburg PA
CBHW021813190326
41518CB00007B/579